YO

YOUNG HAMLET

Essays on Shakespeare's Tragedies

BARBARA EVERETT

CLARENDON PRESS · OXFORD

Oxford University Press, Walton Street, Oxford OX2 6DP
Oxford New York Toronto
Delhi Bombay Calcutta Madras Karachi
Petaling Jaya Singapore Hong Kong Tokyo
Nairobi Dar es Salaam Cape Town
Melbourne Auckland
and associated companies in
Berlin Ibadan

Oxford is a trade mark of Oxford University Press

Published in the United States
by Oxford University Press, New York

© Barbara Everett 1989

First published 1989
First issued as a paperback 1990

All rights reserved. No part of this publication may be reproduced, stored in a retrieval system, or transmitted, in any form or by any means, electronic, mechanical, photocopying, recording, or otherwise, without the prior permission of Oxford University Press

This book is sold subject to the condition that it shall not, by way of trade or otherwise, be lent, re-sold, hired out or otherwise circulated without the publisher's prior consent in any form of binding or cover other than that in which it is published and without a similar condition including this condition being imposed on the subsequent purchaser

British Library Cataloguing in Publication Data
Everett, Barbara
Young Hamlet: essays on Shakespeare's tragedies.
1. Drama in English. Shakespeare, William,
1564–1616. Tragedies.
Critical studies
I. Title
822.313
ISBN 0–19–812254–3 (Pbk)

Library of Congress Cataloging in Publication Data
Everett, Barbara.
Young Hamlet: essays on Shakespeare's tragedies / Barbara Everett.
p. cm. Includes index.
1. Shakespeare, William, 1564–1616—Tragedies. 2. Tragedy.
I. Title
PR2983.E94 1989 822.3'3—dc19 88–34289
ISBN 0–19–812254–3 (Pbk)

Printed and bound in Great Britain by
Biddles Ltd, Guildford and King's Lynn

ACKNOWLEDGEMENTS

The following essays are based on materials which first appeared elsewhere, and I should like to thank Editors for their kind permission to publish them here: '*Hamlet*: Growing' (*London Review of Books*, 31 March 1988); '*Romeo and Juliet*: The Nurse's Story' (*The Critical Quarterly*, Summer 1972); '*Hamlet*: A Time to Die' (*Shakespeare Survey* 30, 1977); 'Textual Readings and Reading the Text of *Hamlet*' (*Review of English Studies* 154, May 1988); 'The Inaction of *Troilus and Cressida*' (*Essays in Criticism* XXXII, April 1982): '"Spanish" Othello: the Making of Shakespeare's Moor' (*Shakespeare Survey* 35, 1982); 'Two Damned Cruces: *Othello* and *Twelfth Night*' (*Review of English Studies* 145, May 1986).

CONTENTS

INTRODUCTORY: Young Hamlet   1

PART ONE
PURCHASING EXPERIENCE

1. *Hamlet*: Growing   11
2. *Othello*: Mixing   35
3. *King Lear*: Loving   59
4. *Macbeth*: Succeeding   83

PART TWO
APPROACHES TO THE TRAGEDIES

5. *Romeo and Juliet*: The Nurse's Story   109
6. *Hamlet*: A Time to Die   124
7. Textual Readings and Reading the Text of *Hamlet*   137
8. The Inaction of *Troilus and Cressida*   165
9. 'Spanish' Othello: The Making of Shakespeare's Moor   186
10. Two Damned Cruces: *Othello* and *Twelfth Night*   208

INDEX   227

Introductory: Young Hamlet

ONCE literary works become classics they may start, paradoxically, to lose their aspect of human truth. In the essays which follow I am concerned mainly with those four of Shakespeare's tragedies most generally thought of as classics: *Hamlet, Othello, King Lear,* and *Macbeth*. My subject is their truth to ordinary human experience. To think of great poetic drama in this way, as if it had been written in our own time, might seem to entail a lack of interest in historical context. But engagement with literature may be more complicated than this, and it can be the very sense of a work's age that brings with it a sharper regaining of its modernity. These questions are not irrelevant to Shakespeare's tragedies, all of them in some sense rooted in history or histories, old stories in new settings—and of none is this more true than of *Hamlet*, Shakespeare's tragedy of fathers and sons, of the past in the present. From it I have taken the phrase 'Young Hamlet', spoken of course by Horatio, to use as title for both this introductory essay and for the collection as a whole, in the hope that it may serve allusively and representatively. When the first scene of Shakespeare's first great tragedy ends by introducing the name of 'Young Hamlet', it tells us at once that the character *is* young, that he embodies all the newness and immediacy of experience itself, but that he is also the past recalled—old Hamlet's royal son and heir.

Hamlet, father and son, live at Elsinore (or so the castle seems to be called): a literary place-name, originally one of the play's crammed but quite minimal details, but now part of the whole *Hamlet* myth. The interesting fact is that the 'true' history of Amleth/Hamlet was acted out in Denmark's main peninsula, Jutland, not in Zeeland (as here) and not in the castle of Kronborg which we have learned to know as Elsinore. Behind Shakespeare's Elsinore there is a curious history which I want to trace briefly in this Introduction, as a way of throwing light on the whole question of *Hamlet*'s modernity, and of what we mean when we say that Hamlet is always young. In Shakespeare's conjectured 'Elsinore' is located all the tragedy's ancientness, its depth of history—but the name and place reflect also the work's peculiar presentness of experience.

In 1600, at around the time when Shakespeare's *Hamlet* had its first performances, King Christian IV of Denmark began to build the great palace he named after his dead father, Frederiksborg. But until it was complete the country's most important royal residence remained the castle Christian had inherited as a boy of eleven when his father died in 1588: Kronborg, on the very edge of the narrow north-eastern Sound between Denmark and Sweden. Known to Danes, then as now, only by this name, the castle was to become familiar elsewhere by another. Down the coast from it stretches Helsingør, a township whose wealth was dependent, from the fifteenth to the nineteenth century, on the Sound Dues claimed by the Danish Crown: in Shakespeare's time the centre of all Anglo-Danish trade, with a sizeable Scottish contingent in it. It is to 'Elsinower' (as the Folio spells it) that Hamlet welcomes Rosencrantz and Guildenstern.

The decision to change the action of *Hamlet* from Jutland to Elsinore may have been taken by the preceding *Hamlet* dramatist; but Shakespeare's world has a strong realism, an awareness of the Europe of his time, which the ur-*Hamlet* is unlikely to have possessed. Certainly there is one simple reason why Shakespeare may have had an alert interest in Denmark and the Danes. Throughout the 1590s James VI of Scotland, to become Shakespeare's patron when he wrote as 'The King's Man', steadily emerged as Elizabeth's probable heir; and his Queen, Anne of Denmark, was Christian's favourite sister (many modern readers take their most vivid image of James's English Court from that farce of drunken masquers occasioned by Christian's visit to his sister in 1606).

But the change to Elsinore in particular, whether instituted by Shakespeare or merely implemented by him, perhaps requires some explanation. While *Hamlet*'s link with the castle of Kronborg is still sometimes honoured by productions of the play which visit it, as travellers of sensibility once did, not much academic interest moves in that direction now; most good modern editions leave the allusion to Elsinore unannotated and undiscussed. The last twenty or thirty years have seen a small revival in scholarly attention, with a series of articles which usefully amass information about the Denmark of Shakespeare's time.[1] Their bearing is mainly political and geographical, and they valuably help to substantiate the image of an Elsinore which was the Gibraltar of the North, commanding the straits through which an enormous amount of trade passed. It is, however, a slightly different aspect of Elsinore and of Kronborg that I want to consider here: an aspect to which the clue is still the name of 'Elsinore' itself.

That the dramatist used, as no Dane would, this Englished name for the castle, the name of the township, suggests that he derived his knowledge at least as much from direct report as from books and maps: that he talked with merchants and especially with actors who had visited Helsingør. And in fact three members of Shakespeare's own company, William Kempe, George Bryan, and Thomas Pope, were almost certainly among the English players who acted there in 1585 and '86. At that date, one topic would have figured large in the conversation of travellers on their return: Kronborg itself. In these precise years the castle was finally transformed from a ruinous old medieval fortress into a great Renaissance palace, blazing with colour and light—indeed, it seems to me probable that the visits of the players were in themselves a part of the festivities organized to celebrate the magnificent completion, after more than a decade, of the palace itself.

Kronborg's history is relevant here. Christian IV, only a child at the time of the 1585-6 celebrations, and his father King Frederik II, were to be remembered historically as

[1] See, for instance, E. A. J. Honigmann, 'The Politics in *Hamlet* and "The World of the Play" ' *Stratford-upon-Avon Studies 5*, (1963), 129-47; Martin Holmes, *The Guns of Elsinore* (1964) 46-53; Gunnar Sjogren, 'Hamlet and the Coronation of Christian IV', *Shakespeare Quarterly*, 16 (1965), 155-60; Keith Brown, 'Hamlet's Place on the Map', *Shakespeare Studies IV* (1968), 160-82.

almost equally possessing certain marked gifts—gifts to some extent shared by many of the rich and ambitious at this period. Celebrated now as Denmark's greatest monarch because of the scale of his attempt to transform his country into a modern European society, Christian realized his aim in one form easy to appreciate now: he built houses. He built (through his armies of masons and painters and weavers) the enormous Frederiksborg and the exquisite Rosenborg and many other great palaces. And he followed his father in rebuilding Kronborg. For Kronborg was primarily King Frederik's achievement. When Frederik married, fairly late in life (in the 1570s), he withdrew from his hitherto militaristic career and set about founding a family. And he built a palace for them. Perhaps because of its vital position, he turned to Krogen, the gloomy and damp and ruinous fortress on the Sound, fortuitously amassed and untouched for a hundred and fifty years. He unified and harmonized, regularized, resurfaced, and above all brought light into a design now single. Within a decade he had transformed Krogen, by teams of gifted and distinguished craftsmen, into a great new Renaissance palace, now called Kronborg. The work was continued for a while by Christian, after he came to the throne in 1596; indeed, the Kronborg which survives is in a sense his, for after the great fire of 1629 Christian rebuilt the whole castle, keeping however so faithfully close to his father's plans that, as early pre-fire illustrations reveal, we still see now much the same Kronborg as Shakespeare's fellow-players saw.

To say this of a structure burned down not long after the dramatist's death is of course to invoke the whole mystery and paradox of tradition, of continuity. But such subjects are hardly irrelevant either to *Hamlet* or to Elsinore or to Kronborg itself. It is interesting, for instance, that two of the pieces of 'local colour' least doubtfully ascribable to the real castle as influencing Shakespeare's drama are sights and sounds such as might most vividly have struck a night-visitor to the interior of the palace on the Sound: the din of trumpets and gunfire which there saluted each royal drinking, and the great tapestries commissioned by Frederik to surround his Great Hall and his Little or Council Chamber, arrases more than large enough to conceal an old prying Councillor himself. But a visiting player might note more than simply the noise, the scale and the splendour. There are things here to be communicated by any exact memory to a consciousness as alert, as intensely receptive as Shakespeare's: and they could be described as a special endowment of the historic, a containment of past in present. The trumpets and guns at the King's drinking were in fact an ancient, even barbaric custom deliberately retained by Frederik in his magnificent new palace, and they were to become known throughout Europe. The great modern tapestries he commissioned (some still survive) also reached back into the past, portraying Saxo's line of Danish royalty, which excludes the usurper Hamlet; a line brought to a climax in the immediate present with the image of Frederik and his young Prince, together in their new Renaissance palace.

Architectural historians now describe Kronborg as spanning three phases: the Gothic, the Renaissance, and the early Baroque, all fused by 1600. It is perhaps permissible to restate this less technically, in more 'human' terms. Kronborg fascinates because it represents a whole historical transition: in it, the Castle has become the Palace, and the Palace the House. I find it hard to believe that returning travellers did not at some time communicate to Shakespeare things that lit up his imagination of what, from our distant perspective, we can call the whole Northern Renaissance. They had witnessed what was for their time the Shock of the New.

'Elsinore' is, in fact, more a human datum than merely an architectural one. A part of Shakespeare's creative genius consists in the degree to which he so embodies a whole human world as to convince the imagination that location too is implied—that everything is implied, and is present before us. But no human truth is dissipated or sacrificed to mere scenery. If Hamlet's welcome to 'Elsinore' is interesting it is for reasons which go beyond what might be called the archaeology of drama, the sediment in it which scholars may helpfully annotate. 'Elsinore' tells us about the whole work of art, and about the character who speaks the welcome.

But this process is in practice two-way and reciprocal—it requires the presence both of the maker and of the audience. If, in the course of the nineteenth century, travellers lost interest in Kronborg, found it un-Shakespearian, one of several possible reasons may have been that they had begun imaginatively to rebuild Elsinore. We may, in short, now give *Hamlet* an architecture it may not have; and that given architecture may span experiences even more than it does buildings. One small example may clarify this. It is very hard to discuss critically or even conversationally the superb first scene of *Hamlet* without including the word 'battlements' (or ramparts, or something of the kind). But whether Elsinore *has* battlements is a difficult question. Kronborg's cannon, as it happens, menace the Sound from a ground-level gun-platform, albeit a raised ground-level: and the gun-platform is in itself almost a 'terrace', such as we might find overlooking the gardens of a great house. This does not argue that Shakespeare himself did not have battlements in mind. He certainly gave his castle an adjacent and towering 'cliff'. But the question seems worth leaving open, if only as a perpetual caveat to the self, that we incline to stage the tragedy, in our minds whether or not in our theatres, in terms that are positively Gothic. And a stringent honesty may insist that we alter the word to Gothick, since we are more likely to work in terms of a late, post-Romantic or post-Victorian, 'current' rather than 'modern', imagining of an Elizabethan imagining of a Middle Age behind it. Moreover, this architectural fantasy (I use the word art-historically, in the sense of *capriccio*) is instinct with a psychology. We give Hamlet's tragedy the buildings we think it deserves, and vice versa. The Olivier film of *Hamlet*—now of course very dated, but in this respect representative of a whole surviving nineteenth-century vision still shaping otherwise good stage productions and critical essays—joined a Tower-of-London massiveness to a factitious cardboard mystery. The blond thickening middle-aged Prince wandered about in a labyrinthine and gloomy pile as the 'man who could not make up his mind', his intellectual stasis merging into the winding staircases of the multifarious ruin where he lost himself. The conceit dates back to an apprehension of the tragedy wittily summed up by Mallarmé's Prince who walks reading the book of himself, 'Il se promène, lisant au livre de lui-même'; and it was most resonantly Englished in the castle of Eliot's 'Gerontion', History's 'cunning passages, contrived corridors'.

One of the hardest but most fascinating of all intellectual problems is how not to patronize the past. We err, certainly, by forgetting that the past is not the same as ourselves. But we may err even more by not remembering that the past probably has more in common with us than we think. Gothicizing and Romanticizing are ways of keeping things in their place, of pleasurable and affectionate patronage. For Tudor and Gothic (even leaving the Gothick out of account) are not the same; in the 1590s the Middle Ages were so lost in the mists of antiquity that jousts could be revived in Elizabeth's Court for the charm of the far distant. The tragedy of *Hamlet* involves a knife-edge sense of modernity, almost that experience of the present instant between gulfs of past and future which in *Troilus and Cressida* becomes a vertigo. The palace of Kronborg at Elsinore turned a comparable experience of past, present and future into an architectural marvel for travellers to talk about.

The sense of the past that is deep in *Hamlet* we are perhaps unlikely to make a mistake about, partly because the play is so rooted in our own cultural past, and partly because the past actually walks in it, in the character of the Ghost. There may be more need for criticism to think more questioningly of this and the other tragedies' peculiar modernity. Concepts of the 'historic' and the 'modern' are so complex, even so metaphysical, that simplifications are useful—and even in the aesthetic there is room for the utilitarian. I spoke of Kronborg earlier as a transition from the fortress to the palace, and from the palace to the house. In the simplest terms, a house is a place where a family lives. Frederik began building Kronborg out of Krogen, the old military citadel, only after his marriage, and it was the birth of his son and heir Christian which in 1577 brought the enterprise to its first climax in festivities.

Certainly, Shakespeare's tragedies have to do with the 'fortress'—they are activated by the great battles of power-politics. But, if the dramatist far excelled his contemporaries, it was partly by virtue of his understanding how deep these issues of power went in the individual, and how deeply they were transformed. The fortress becomes the palace, and the palace the house. *Hamlet* is the story of a son who must—as the young always must—by living accept an inheritance largely unwanted from the generation of the fathers. It communicates to us now because our age is, like Shakespeare's, moving frighteningly fast from the past into the future. The one stable 'Elsinore' is the individual consciousness as it recognizes and responds to that shift. Within the 'house' of civilization, within the 'family' of the race, the most basic and natural expression of that process is the inevitable growing and growing-up of the young: the tragedy of Young Hamlet.

In each of the four essays which follow this Introduction I have hoped to give some account of the sense in which Shakespeare's great tragedies are all about these basic, even elemental aspects of human experience. These four chapters first took the form of the Lord Northcliffe Lectures, delivered at University College London in 1988, under the general title of 'Purchasing Experience'. I was delighted and honoured to be invited to give these lectures; and I owe special thanks to Professor Karl Miller for encouragement and for one item of vital help in particular. To this first group of essays I have added, in Part Two, some attempts to offer solutions to specific problems in two of these tragedies, in the early tragedy *Romeo and Juliet*, and in the near-tragedy *Troilus and Cressida*. My very grateful thanks are due to Professor Emrys Jones for help given throughout the book.

PART ONE
Purchasing Experience

'... unlesse Experience be a Jewell, that I have purchased at an infinite rate.'

The Merry Wives of Windsor, II. ii. 198–9

1
Hamlet: Growing

BBC Radio has started a pleasant practice of filling the Christmas season with murder plays, mostly dramatized detective stories from the classic English phase of the 1920s and 1930s. This joining of the festive with the lethal provokes thought. There may well be some long line in English culture that links the Christmas visit to *The Mouse Trap* with a point at least as far back as that splendid moment in medieval literature when the Green Knight, his head cut off, stoops to pick up the rolling object, and rides out of Arthur's Christmas Court with the head lifted high and turned in the hand to smile genially here and there at the gathered knights and ladies as he goes. 'A sad tale's best for winter'. If there is such a tradition of smiling violence, then clearly there must be a place in it for the original 'Mouse Trap' itself, Shakespeare's tragedy of Court life. Indeed, as the work of the most formally inventive of all literary geniuses, *Hamlet* could even be called—particularly since its presumed Kydian predecessor is lost—the first ever detective story or civilized thriller. The drama critic James Agate, who once savagely described Donald Wolfit's Hamlet as a private detective watching the jewels at the Claudius–Gertrude wedding feast, may have said more than he knew.

Yet to praise *Hamlet* as the first detective story makes sense mainly in terms of a conceit, feasible partly because ridiculous. Literary artists have worked in the genre: Poe, Wilkie Collins, Simenon, Chandler, and Michael Innes among others. But the true English 'classics' of the 1920s and '30s, the books we evoke in recalling a body in a locked library in a country house, hardly go in for artistry. V. S. Pritchett once wrote down the whole genre as Philistine, and many are notably badly written, their characters a stereotype and their language a cliché. These particular classics of the 1930s are fictions that evade not only the public horrors they seem faintly to shadow, but more private intensities of self-contemplation: they work in short as ritual games and puzzles, effective by their exclusions—their interesting mix of violence and nullity the opium, perhaps, of the English rectory and manor-house during the troubled inter-war period.

With an extraordinary unanimity, good representative Introductions to *Hamlet* and critical essays on it speak of it as the most enigmatic play in the canon, 'the most problematic play ever written by Shakespeare or any other playwright':[1] a theme that has been with us since the late eighteenth century or thereabouts, when the tragedy for the first time began to become 'a mystery', a 'question', most of all a 'problem'. Some critics assume that for all its interest *Hamlet* simply fails to hold together; others more tentatively voice the difficulty of knowing 'what the play is really about'.[2] There can be a kind of relief in recognizing that the problematic may itself be subdued into entertainment—that 'Whodunit?' is in itself a pleasurable system, with the Prince as detective, victim, and villain rolled into one. Acted out in a confined world of rituals and conventions—court politics, revenge, a clock that records time always passing—the tragedy of *Hamlet* gives the deep if quizzical solace of all games and puzzles. Its hero an undergraduate, the dominance in the play of a great freewheeling exercised intelligence ('I will walke heere in the Hall . . . 'tis the breathing time of day with me') makes the work what it is, the world's most sheerly entertaining tragedy, the cleverest, perhaps even the funniest: Dr Johnson meant this when he gave it 'the praise of variety', adding 'The pretended madness of Hamlet causes much mirth'.

Yet even Johnson had something like 'problems' with the

---

[1] Harry Levin, *Shakespeare Quarterly*, 7 (1956), 105. Both this and 'what the play is really about' (see n. 2 below) are quoted in the Introduction to the New Arden *Hamlet* (1982; ed. Harold Jenkins).
[2] A. J. A. Waldock, *Hamlet: A Study in Critical Method* (1931), 7.

play: he was too honest not to mention that 'Of the feigned madness of Hamlet there appears no adequate cause'—for the very distinction of *Hamlet* is the degree to which it makes us reach out and in for 'causes'. The tragedy is, even when compared with those earlier dramas of Shakespeare which have done so much to nourish it, the English Histories in particular, self-evidently great. But this is not *just* because it is entertaining: or rather, it has managed to stay so for four hundred years because the human mind, which is entertainable in a large variety of ways, is always discovering 'a hunger in itself to be more serious'. The Christmassy detective stories of the Twenties and Thirties we are nostalgically reviving now precisely for their unseriousness, the efficiency with which they don't matter. Shakespeare's tragedy matters. It means something.

Hamlet is sometimes described as the first great tragedy in Europe for two thousand years. The achievement perhaps owes something to Shakespeare's unique mastery of ends hard not to state as opposites: the power to entertain and the power to mean. The bridging of the two in the writer's work gives some sense of his giant reticent power of mind. To quote William Empson, on another subject, 'The contradictions cover such a range'—yet they are always reciprocal, in communication with each other. The meaning of *Hamlet* must be intrinsic with what in it holds audiences and readers. And, even if *King Lear* has come, at this point in a violent and in some sense paranoiac culture, to be ranked highest as 'the greatest', *Hamlet* is still manifestly enjoyed, indeed loved by readers and audiences, even by critics. It has always been so, ever since the extraordinary furore caused by its early performances, probably in the last year of the sixteenth century. The tragedy seems to have been one of the world's great successes, producing—we can tell from the profusion of admiring, amused and envious contemporary references, quotations and parodies—a kind of matching madness in its audiences. It may be that this was the response of human beings to a literary work that went deeper, not just entertainingly wider but truthfully deeper, than any aesthetic work of their experience: deep in a way that was slightly out of their control.

There is an old joke about *Hamlet* being full of quotations. So it is; but perhaps so it always was, even for its first audiences. If we can associate this great drama with detective stories and thrillers trivial at their best, as a 'sad tale . . . best for winter', the reason is that 'Christmassy' quality I have tried in passing to hint at: something in the drama profoundly reminiscential, nostalgic, obscurely looking back, so that the murderous Court and castle are none the less eerily cosy, as if we had always lived there. This familiarity may be explained on one level by the fact that the tragedy is so deep-rooted in our literature, our culture, even our schooling; and its first audiences perhaps savoured in much the same way the work's relation with a now non-existent predecessor. But this harking-back quality is intrinsic, not merely incidental. The play's opening scene builds a great past for the work to inhabit; even the Ghost has been before, last night and the night before that. Recurrence has its climax at the end of the scene, which features an actual mention of Christmas, 'that Season . . . | Wherein our Saviour's Birth is celebrated'. The play would be different without this curiously stirring legend that 'The Bird of Dawning singeth all night long'—it would lack some endorsement, some sense of otherwhere. Commentators don't seem to ask why Shakespeare failed to use the more conventional association of cockcrowing with Easter. But the New Arden editor is surely right to hint that the writer invented rather than found this myth of Christmas. And if Shakespeare did so, then his reason was that Christmas was needed in his play.

One of the great Christmas texts is Isaiah's 'Unto us a child is born, unto us a son is given'. Horatio answers Marcellus's Christmas speech with

> Let us impart what we have seene to night
> Unto yong *Hamlet*.[3]

---

[3] Shakespeare quotations are from *The Norton Facsimile: The First Folio of Shakespeare* (1968), prepared by Charlton Hinman.

Coming as this does so soon after the visitation by the late King's Ghost, this designates Hamlet as the Son, the Prince, Hamlet Junior. But the phrase 'Young Hamlet' has a more absolute meaning. In an interesting early allusion, the writer of an elegy for Richard Burbage after his death in 1619 names the actor's great roles as

> young Hamlett, ould Hieronymoe,
> Kind Leer, the Greved Moore[4]

—where Hamlet is young as Lear is kind and the Moor grieved. The phrase, which may have been regular in use, gives a valuable suggestion as to something vital in the tragedy that we have now largely lost.

Artists may work through a highly sophisticated complexity —such as Shakespeare certainly mastered in his career—to achieve a startlingly original simplicity. That simplicity comes into being with the existence of 'Young' Hamlet. The Court that holds him is a brilliant creation reflecting back to us high-level existence of the period in all its details, both mundane and powerfully formal: the 'World' as Shakespeare perceived it at this height of his career. And it is this Court which makes the tragedy so real, so permanently interesting, so unshakeably ambiguous. But we see it through a given very individual observer, an innocence, however corruptible, locked inside its experience: Young Hamlet.

Our tragic sense is mainly inherited from two sets of ancestors: the universalizing Victorians and the symbolizing Modernists. Both in their antithetical ways may equally neglect simple facts about Elizabethan life and art. The only great tragedy we have derives from a materialistic culture whose philosophy is obsessed by Crime and Punishment. Its highbrow reading is characterized by Seneca's closet-drama, whose dingy gang-warfare of revenge bequeathes to *Hamlet* itself its local claustrophobia. The two brilliantly gifted predecessors to whom Shakespeare perhaps owed most, Marlowe and Kyd, were both in their very different ways

---

[4] *The Shakspere Allusion Book* (1909), 272.

absorbed by this same mechanism of sheer human will, Marlowe through the great thugs who became his heroes and Kyd through his insight into the Machiavellian Court.

What is original to Shakespeare is his revelatory sense of the natural, of what is both fresh and classic in human feeling and human experience. In a now rather underrated comedy, *The Merry Wives of Windsor*, the burgher Master Ford thinks of experience as a jewel he has purchased 'at an infinite rate'.[5] But purchasing experience is what all Shakespeare's characters are in the end engrossed by. And what manifests the writer's original genius from the beginning is his unique sense of what makes experience fully human, shareable yet monolithic in its elements. In his very first two tragedies, Titus Andronicus is a father, and Romeo and Juliet are children. The long, rich series of what may be formally Shakespeare's own invention, the English History Plays, work in *Henry IV*, Parts I and II to a climax that could be subtitled *Fathers and Sons*.

Implicit in these Histories from the first is a tragic potential: the weak Henry VI must observe the effect of his weakness, a son who has killed his father, and a father who has killed his son. Only in *Hamlet* does that potential fulfil and complete itself, as the play gives to history dimensions at once mythical and internal. Its soliloquizing hero is the Prince, the Son, the 'young 'un': his remembering consciousness is a surrogate for, almost at moments a part of the past of, the play's readers and audiences, reflecting as they watch a drama itself holding 'a mirror up to nature'. Young Hamlet, as it were all Europe's 'Elder Son', the white hope of history, grows up to find that he has grown dead: his is the body in the library not merely of Elsinore but of Western culture at large.

Victorian literary critics sometimes asked questions whose literalism makes them equally wrong-headed and useful. The practice is summarized in the mocking footnote we attach to

---

[5] *The Merry Wives of Windsor*, II. ii. 216–17.

Bradley, 'How Many Children Had Lady Macbeth?' A similar topic once much debated though now rarely reverted to is: 'How old, exactly, is Hamlet?' The dimensions of the problem are these. The Prince is introduced to us as an undergraduate. But in the last phase of the tragedy, when Hamlet is returning home after long absence on the high seas, he meets in the graveyard a Clown who tells him how old he must be, by dating his own long career as gravedigger:

> I came too't that day that our last King *Hamlet* o'recame *Fortinbras*
> ... It was the very day, that young *Hamlet* was borne, hee that was mad, and sent into England ... I have bin sixeteene [sexton] heere, man and Boy thirty yeares. (v. i. 159–77)

The undergraduate is therefore thirty years old. Scholars and commentators once used to try to resolve these figures by dissolving them. But this is a mistake. The 'thirty' is confirmed by Hamlet's own contribution to the chain of figures, his evocation of his own childish self on Yorick's shoulders—and Yorick has, we are told, been dead just twenty-three years. Equal solidity is the rule early in the play. Shakespeare has worked at Hamlet's establishment as an undergraduate, and the confusions of Horatio's role may in part depend on it. Moreover, those haunting repetitions in the Prince's speech which occur only in the Folio ('Indeed, indeed sirs'; 'Wormwood, wormwood') and which the New Arden attributes to actors' interference may rather be, in my opinion, Shakespeare's additions to help characterize the undergraduate by making him do what the type still touchingly does: imitate the pedantic mannerisms of an admired tutor.

Why should Shakespeare so desire to make real Hamlet's attachment to Wittenberg? The place had of course its own connotations—it was Faustus's university, as well as one much attended then by Danes of good birth. But Wittenberg is surely no more than likely; what is necessary is making Hamlet an undergraduate. And it is necessary because it is the simplest, solidest but most economical way of showing the Prince as young. Further, although Hamlet's age in years is not the important issue, as an undergraduate he would by no means be thirty years old. The play takes us directly and deeply at many points into a social history whose loss can cut off from us a sense of the work's simplicities. A university education at this almost incomparably well-educated period was too important, yet also too common a factor for a dramatist to play games with, particularly given that most of Shakespeare's literary contemporaries and a good number of his audience had been to Oxford or Cambridge or the Inns of Court, or some combination of these. If Hamlet went to the university, he was between about sixteen and about twenty-three, the seven years that allowed for first the BA then, for those who stayed on, the MA.[6] Aristocrats and Roman Catholics went to the university earlier, sometimes much earlier; only eccentrics like Gabriel Harvey stayed for ever.

These are the norms, and Hamlet's intellectual youth, high in nuisance value, indicates that he adhered to these norms. Even the black of his mourning garments must have helped suggestively to support the point of his youth (attention is drawn to it): for university 'subfusc' was so much more intensive at this time, and so unavoidable by the young, that black indicated the scholar as much as the bereaved.[7] The peculiarly social, even legal role of academic black, the part it played in holding certain youthful orders in thrall, serves as a useful key to a larger question. There are issues here not at all easy to disentangle, and both *Hamlet* and Elizabethan society in general positively refuse to keep them apart. I have hinted that to ask precisely how old Hamlet was has to be classed as a non-question. Many Elizabethans would have agreed in regarding it as a non-question, partly because—as Keith Thomas has pointed out in an admirable lecture on 'Age and Authority in Early Modern England'—many didn't themselves know how old they were. What they did know, or

[6] Lawrence Stone, 'The Educational Revolution in England, 1560–1640', *Past and Present*, 28 (1964), 57.
[7] See, e.g., John Earle, *Micro-cosmographie* (1628), No. 24, 'A Meere young Gentleman of the University': 'Of all things he endures not to be mistaken for a Scholler, and hates a black suit though it be of Sattin.'

resentfully wished not to know, was their place—and youth had, or resentfully wished not to have, a place in sixteenth-century society. This issue even has for historians its special focus in terms of undergraduates:

> In the history of the universities in England the late 16th century stands out as the age when a young man's 'university days' first came to be regarded as a period for the 'sowing of wild oats' . . . A gentleman of Renaissance England fitted himself for his future role of governour . . . *after* he left university.[8]

The university is the characteristic ante-room or waiting-place for the life of mature years: the life of power.

'Age' in Renaissance Europe is politically and socially adjusted. Behind that adjustment is a very thorough negation of modern systems of timing and ageing. There are in short figures and numbers in Shakespeare's tragedies, and they matter, but they are not our figures and numbers. Recent studies from both social and literary historians have documented that vague sense which any reader can get from Elizabethan literature that its characters simply don't know how old they are. The most complete and compendious of these scholarly works, John Burrow's *The Ages of Man*, reaches back through medieval to classical times to show how very differently existence was measured before our own pervasive if shallow mathematical and technological revolution. He reminds us that the Gospels record no fact in Jesus's life between his boyhood encounter with the Doctors at twelve, and the beginning of his ministry at thirty or thereabouts. This pair of dates blended with the classical, principally Aristotelian, patterning of human life into what were sometimes seven but more often three stages: the loose and variable but influential categories of 'youth', 'maturity', and 'age'. The dates of these stages can shift considerably; their concepts cannot. This is the 'positive' Burrow argues. His book contains a 'negative' which seems to me of equal importance:

---

[8] Kenneth Charlton, *Education in Renaissance England* (1965), 150.

Most ancient and mediaeval authorities speak of the course of human life not as a process of continuous development but as a series of transits from one distinct stage to another . . . They generally saw the transitions between these estates as datable events rather than gradual processes. Mediaeval narrative displays a corresponding lack of interest in the process of change from one age to another . . .[9]

An example Burrow gives is the reminder that 'the events of Chaucer's Knight's Tale occupy more than a decade', but that 'Palamon and Emily remain "young" throughout'. *Hamlet*, too, shows signs of spanning something like a decade—though we should be unwise to time the play on these terms. But Hamlet does *not* 'remain "young" throughout', unlike any medieval or indeed contemporary character. Shakespeare is imagining that procedure of development and continuity, at once internal and psychological yet locked up in public events, which seems simply non-existent in most earlier thinking. But to appreciate this we have to 'set back' our clocks, and—as when crossing some geographical date-line—allow for a very different time-system.

The vital question of development or 'growing' I shall return to. For the moment I want only to establish that the notion of time most relevant here is a definition not merely in terms of human life and human procedures, but specifically political. Age is status, age is power. Or rather, maturity is status and power: for in this period, so Keith Thomas argues, age recedes to join youth among the disadvantaged:

In early modern England the prevailing ideal was gerontocratic: the young were to serve and the old were to rule . . . By analogy it justified the whole social order; for the lower classes at home, like the savages abroad, were often seen as 'childish' creatures, living in a state of arrested development, needing the mature rule of their superiors.[10]

But, Thomas points out, the sixteenth and seventeenth

[9] John Burrow, *The Ages of Man* (Oxford, 1986), 177–8.
[10] Keith Thomas, 'Age and Authority in Early Modern England', *Proceedings of the British Academy*, LXII (1976), 207–10.

centuries—except when social crises like war enforced otherwise—defined 'maturity' so narrowly within the forty- and fifty-year-olds, as to create in practice an enormous class (over 90 per cent) of the 'young' and the 'very old' who were disadvantaged and dependent.

It may not be irrelevant to remember the main characters of the first three of Shakespeare's great tragedies: Hamlet, whose youth disadvantages him; Lear, whose extreme age debilitates him; and Othello, whose Moorish distinctiveness allows the vicious to alienate him as a 'stranger' within his own society, so undermining his power and apparent maturity. Macbeth like Eliot's Gerontion seems to have relationship with all three human estates, youth, maturity, and age, and to forfeit the rewards of all three by the damage he does to his own humanity. Of the first three, Hamlet, Othello, and Lear, the young man, the 'alien' mature man, and the old man, we must surely feel that these choices of character are not accidental. Shakespeare is spanning the full range of adult experience. Yet to their disadvantage or capacity for suffering the writer adds a complicating factor: each of these heroes is royal, or has some relation to royalty. The fact that Hamlet is the Prince, in a play that carefully obscures Denmark's politics of primogeniture, gives a clue to what Shakespeare has done with the role. Hamlet is all the double strength *and* weakness of youth and sonship; and this always ambiguous state of youth, held in the memory of every grown member of Shakespeare's audience ('you your selfe Sir, should be as old as I am, if like a Crab you could go backward'—'going backward' is at once remembering, retrogressing, and making obeisance to royalty) in its turn generates a large and wholly original tragedy of consciousness.

I am suggesting that the unusual degree of political vulnerability attending the out-of-power 'young' in Shakespeare's England helped him towards the creative ambiguities locked up in his first great tragic hero. Three hundred years after Shakespeare, a writer possibly influenced by *Hamlet* in this, Henry James, made a fascinating near-tragedy out of the subject of the 'Awkward Age', the period in the lives of young late-Victorian women when they were neither in the schoolroom nor entirely out in the world, safely married—a period of cruel subservience and limitation yet also of heroic freedom to see and to think. Because of the political conditioning of 'youth' in the English Renaissance, Shakespeare's Prince himself exists at an 'Age' or time which the dramatist has invented, a time also savagely confined yet endowed with peculiar freedoms. To grasp this time which is the heroic medium is perhaps to resolve into simplicity some of the elements in the drama which have become its most inordinately examined problems. I have already mentioned Dr Johnson's protest that though he found it entertaining he saw no dramatic 'cause' for Hamlet's 'madness'. But he had perhaps forgotten what he once said (or Boswell at least makes him say) of his own undergraduate youth at Oxford: 'Ah, Sir, I was mad and violent. It was bitterness which they mistook for frolic . . .'[11] Hamlet too, and many real young people after him, might have added that like Johnson he 'thought to fight his way by his literature and his wit', and that he 'disregarded all power and all authority'. Hamlet's dangerous subversive humour—which is neither madness nor sanity, but a denial of the authority of the society that holds him—permanently defines a freedom and impotence of the young. The same kind of simplifying process may be used for some illumination on literature's best-known line and indeed the whole soliloquy that follows it, 'To be or not to be': long debated as to whether its subject is suicide, or revenge, or any other topic positively. There is a certain relevance in a line from a Philip Larkin poem, '*Vers de Société*': 'Only the young can be alone freely.' Only the young can so detachedly if tormentedly survey the prospect of adult existence as to believe that they have the option 'To

---

[11] *Boswell's Life of Johnson*, ed. G. B. Hill and L. F. Powell (1934), i. 73–4. Dr Keith Walker has kindly mentioned to me Marshall Waingrow's opinion (*The Correspondence and Other Papers of James Boswell* (1969), 57 n. 10) that Boswell misread the 'rude' of his notes as 'mad'. I suspect, however, that Boswell may here as elsewhere in the *Life* have improved on his first thoughts. 'Mad' could be used during the whole period 1500–1800 to mean 'foolish, unwise, extravagant in gaiety'.

be or not to be'; the adult, with 'promises to keep', more often has to shrug and trudge on.

Some part of this extreme originality of insight clearly came to Shakespeare from the Elizabethan political situation—from the dramatist's ability to render private and internal what began in public life. The Jacobean Hamlet, who drove audiences mad with pleasure, amusement and alarm, was a Malcontent, a political subversive—as indeed was the play's Hamlet, being from the first too much the 'son', over-fathered by too much (and too unkind) 'kin'; and so were all those who felt that their society made of them no more than a 'captive good attending Captain Ill'. The new political studies of Shakespeare in general take the stance of assuming that the chief of the King's Men must have been conservative. But the writer of the Sonnets can envisage that 'Captains' might or must be 'Ill'; and poetry itself was in Elizabethan society specifically no more than a toy for the young. The mature man, the man of power, had his attention engaged elsewhere. Thus, 'Authors' (E. H. Miller tells us, in his valuable study of *The Professional Writer in Elizabethan England*) 'viewed themselves as prodigal sons'. Fundamentally, Miller points out, 'Elizabethan writers accepted the premises as well as the fears and aspirations of Tudor culture. "Profite", or utility, was an obsession with English humanists',[12] who therefore speak through Claudius's rebuke to Hamlet for the frank wastefulness, the sheer childishness of his grief for his dead father. Indeed, formally most Tudor writers themselves spoke with him, apologizing for what they tended to throw away (like Puttenham) as 'but the studie of my yonger yeares in which vanitie raigned'.[13] The rest of the world concurred. 'Though they be reasonable wittie and well don yet' (a letter of John Chamberlain's groans over elegiac lines by Donne when Dean of St Paul's) 'I

---

[12] E. H. Miller, *The Professional Writer in Elizabethan England* (1959), 17-18.
[13] George Puttenham, *The Arte of English Poesie*, ed. G. D. Wilcock and A. Walker (1936), 308.

could wish a man of his yeares and place to give over versifieng'.[14]

The point here is in the 'yeares' and 'place'. Even a social and cultural theory most repressive to the young could grudgingly accept the kind of stress laid in Cicero's *Pro Caelio* on the necessary element of 'ludus' to be allowed to the very young: 'Everybody agrees in allowing youth a little fun . . . Let some fun be granted to youth.' With this near-identification of 'youth' with 'a little fun', or in a word '*play*'—with which most Elizabethan educationalists and politicians would have been in nominal accord—we come directly and fully back into Shakespeare's play. I have been hoping to suggest the very large and significant network of conditions—some contemporary, some much more permanent and lasting—'Young Hamlet' brings with him into his Court and his play. It is hard to imagine any of the later heroes as what the Prince is, an amateur but devout poet and playwright and even producer; and Shakespeare plainly gets a good deal of quiet ambiguous pleasure from dramatizing the lordly certainty with which his brilliant young aristocrat tells the tired polite professionals how to act. This amusing detail helps to illustrate the point at which I began, the peculiar entertainingness to be found in the tragedy. From his first wild, dangerous but liberating evasions of Claudius, the ironies that are also (politically) his first 'madness' of youth ('More then kin, and lesse then kinde'; 'I am too much i'th'Sun'), Hamlet brings into the confining world of power exercised that spaciousness of intellectual play which gives the work its largeness. It is significant, in the context, that Claudius first addresses him with a 'But'—'But now my Cosin *Hamlet*, and my Sonne?'—a remark which accompanies the turn from the easily-handled Laertes with a certain casual offensiveness, and Hamlet repays the insult with muffled interest. But Hamlet's own father, loved as he is, reacts to his son with something like the irritation of his hated brother-

---

[14] *The Letters of John Chamberlain*, ed. N. E. McClure (Philadelphia, 1939), ii. 613.

King; the quick true skittering sympathy proper to Hamlet's age and type, 'Alas poore Ghost', the Ghost himself crushingly repels: 'Pitty me not, but lend thy serious hearing | To what I shall unfold.' Both Father-Kings identify the 'serious' with the purposive. From his first skirmish of wit to the final Court-duel that kills him, Hamlet defines the 'serious' so as to contain a violent play, his dance around the two revenge-held Kings opening out in the drama those dimensions and depths which we associate in the comedies with women and Fools, in the preceding Histories with such characters as the Bastard Faulconbridge and the great 'player', Falstaff. But the game gives the Kings and the play itself both time and occasion to destroy him—or Hamlet time to destroy himself.

Hamlet's involvement with 'play' in the first half of the action has its climax, naturally enough, in the arrival of the Players—the more usual Elizabethan word for what we now describe as 'actors'. The Players are often discussed as one of the tragedy's major problems, the randomness of their presence characterizing all in the drama that is wayward and enigmatic. There is of course something in this; Hamlet's whole world is, by accident and by principle, wayward and enigmatic. Yet there is at the same time a very marked decorum in the Players, an appropriateness in their appearance. Two important circumstances seem never to be noted. The Players are carefully introduced by Rosencrantz and Guildenstern, who make them indeed appear to 'hold the mirror up to nature': for their case both resembles and reverses Hamlet's. Where his youth is trapped by Authority in a humiliating close Court, they have been driven out of the city by triumphant children, who—Hamlet at once perceives the relevance—'Exclaim against their own succession', verbally war against the stage of life they must themselves in time come to. The point is underlined immediately as the adult Players arrive, for Hamlet greets them affectionately as the only-just-not-children-themselves that they are: this young face so newly bearded, that boy-girl player suddenly grown tall.

The Players represent what both Kings would no doubt see (Polonius's enthusiasm is depressing in its patronage) as evasion, unseriousness, irrelevance—and in one sense the Kings are right. But Hamlet's comparable 'marginality' also leaves him free to look at things. And what he sees in the Players is not merely a capacity to act 'revenge' entertainingly; he also sees what they are or embody, beyond what they act: a struggle or conflict in human existence that is deeper and more permanent than the revenge-system which it resembles. The helpless division of the generations can lead to mutual destruction. Yet in the understanding of every observant human individual, the generations are really one and the same, mere 'stages' held in one by memory and sympathy,—as Young Hamlet, watching the Player weep for long-dead Hecuba, becomes an audience, crossing a great gulf to join us in the present.

The play scene (III. iii) has become one of the tragedy's great problems, in that it seems to tell us nothing of what Claudius sees in the play-within-the-play, and therefore tells us nothing to the point. But what the play scene does tell us is that 'points' are of more than one kind, as Age and Youth are two different generations. Hamlet's delight in his play, which to him spells the truth, emerges in his succeeding near-intoxication. As readers and audiences we are in no position to belittle his enchantment; the tragedy begins and ends with first Barnardo and last Horatio rendering all its past into the terms of hypnotic story, and while the play lasts it invents those terms by which reader and audience alike become 'young' again, cut off from Time, exchanging active power for a freedom more detached and contemplative.

But the murderous King is hardly a natural playgoer. If he were capable of accepting this kind of truth he would be less successful at his own. He lives by a different clock: 'That we would do, we should do when we would.' Truth as he sees it is not what happened to his brother in some theoretical past but what happens to himself in the real present. 'The Mouse Trap' shows that a King may be killed by his nephew—a term that to Elizabethans could mean a bastard or a grandson or a successor: possibility enough to indicate that the King's stepson was getting far too much of a nuisance:

> Our estate may not endure
> Hazard so dangerous as doth hourely grow
> Out of his bourds[15]
>
> (III. iii. 5–7)

—a theme precisely echoed by Polonius's 'Tell him his prankes have been too broad to beare with'. 'Bourds' and 'prankes' nicely focus the Court insistence on the menace of Hamlet's youth.

It may even be Hamlet's youth which prevents him from killing the King at prayer, for reasons he gives with a savagery partly dependent on frustration. A theoretician, a perfectionist, not yet habituated to brutality, he has accepted Revenge as a system antithetical to most of himself which must be carried out with all the more Old Testament exactitude of eye-for-eye and tooth-for-tooth. To stab on other terms would be dishonouring, a mere act of murder— and Young Hamlet is as much an aristocrat as his royal father. His rage at the indecencies and injustices of mere life carries him straight to the fury of the scene in his mother's bedroom and to the accidental killing of Polonius. From this point of randomness at the centre of the play, clumsy life's mirror-image of the murder recalled by the fiction of 'The Mouse Trap', Hamlet is done for. Claudius corrupts the young and courtly-conventional Laertes and they unite against him. The Avenger has become the object of Revenge.

Readers and audiences of *Hamlet* seem to have experienced remarkably little difficulty with the play until a point late in the eighteenth century. Then, though as ever much loved, the drama began to be found a 'mystery', an 'enigma', and at last a 'problem'. To understand the importance of 'Young Hamlet' may be a way of grasping this interesting critical

---

[15] I here emend F's 'Lunacies' (itself an apparent compositorial emendation of Q2's *brows*) to 'bourds', for which a case is made in a later essay (see pp. 159–61 below).

shift. For literature itself records for us what happened to the reader's image of the tragedy in this period. During the last two decades of the eighteenth century, while in England Wordsworth was living through the materials of his future poem, on the 'Growth of a Poet's Mind', in Germany Goethe was putting together a long, loose yet oddly powerful and even spell-binding romance, which recounted the adventures of a young bourgeois who leaves home to join a troupe of wandering actors, hoping to advance what he believes to be his calling as a writer, an artist: but in the end the hero changes his mind, and goes back to real life with a more practical determination to serve humanity. Renowned through Europe for a century or more, Goethe's story *Wilhelm Meister* is hardly now a current classic, though still saluted by literary historians as the first of the enormously influential literary mode which it invented: the *Bildungsroman*, or story of the life-education of a young person, whose growing-up is specifically defined as a turning-away from a self-centred 'artistic' existence and a committed entry into society at large.

The 'action' of *Wilhelm Meister* is really an intense and sustained brooding on the work which obsesses the young writer Wilhelm—Shakespeare's *Hamlet*, a production of which forms the culmination of the story. The English tragedy becomes something close to a play-within-the-play, an obscure yet suggestive paradigm of the new, organically free, always-emergent life of Goethe's own novel—yet it has this fertilizing power despite or perhaps because of the fact that Wilhelm cannot understand the tragedy, which seems to him a work 'full of plan' that holds at its centre a hero 'without a plan' (the phrases are his own). This paradox assumes considerable significance, if we recall that this is the exact moment when the tragedy begins to become a 'mystery'. Moreover, Goethe is hardly exceptional. After a century of English novels of which the most distinguished are plainly in the line of the *Bildungsroman*, the use made of *Hamlet* by the arch-Romantic writer Goethe is strikingly paralleled in the work of the arch-Modernist Joyce. James Joyce, who knew and (with reservations) admired *Wilhelm*

Meister and even owned a copy, includes in *Ulysses*, his great tragicomic story of a 'Father' and 'Son' reconciled, a long Goethean library-discussion of the meaning of *Hamlet* and its possible provenance in Shakespeare's own life.

The enormous, rich and various inheritance of later English literature from Shakespeare's first great tragedy is not my subject here. One brief point may be made about some of the very best nineteenth-century writing in that area. Though the post-Romantic novel is of course strongly rooted in the *Bildungsroman*, the 'story of life-education', the English imagination at its best turns aside from the German form and intuitively harks back to the Shakespearian ancestor, more at home with the dark or tragic than with the optimistic, socially orientated and progressive Romantic fiction. We ought perhaps to explain the unusual formality and aesthetic coherence of *Great Expectations*, for instance, in terms of the book's being a kind of reverie on the work which actually appears in its thirty-first chapter as Mr Wopsle's *Hamlet*: a reverie which brought into the novel the Christmas graveyard, the dead children, the ritualistic games of 'Beggar-my-Neighbour' in the 'Court' of Satis House, the cruel Petrarchan heroine, the lawyers of Little England, and above all the terrible returner from the dead who is 'your second father. You're my son':—everything in the book, in fact, which seems to say to the reader, with an unnerving dark cosiness of memory, what the last chapters say to Pip: 'Do not thou go home, let him not go home, let us not go home . . .' And one might add to Dickens the very different but almost equally rich case of Henry James, whose story of a father-destroyed and haunted youth of military family, *Owen Wingrave*, reads like a late-Victorian paraphrase of *Hamlet*. I have already suggested that James owed something to Shakespeare when he began his late studies in isolated heroic consciousness with novels and stories whose protagonists were not merely young, but sometimes children. None ends happily.[16]

[16] I explore this point further in 'Henry James's Children', in *The Child and the Book* (Oxford, 1989), ed. Gillian Avery and Julia Briggs.

My point in introducing *Wilhelm Meister* was to propose the dependence of that major form of the European novel, the *Bildungsroman*, on Shakespeare's tragedy. *Hamlet*, the first great story in Europe of a young man growing up, in a sense originates the *Bildungsroman* itself. But if *Hamlet* invents or inspires the form, it also denies it. Romantic culture, taking its stand on the revolutionary hope of youth's great growing-stages, found itself forestalled by Shakespeare in more ways than one. And if the play begins at this period to become a 'mystery', an 'enigma', a 'problem', the reason surely is what the latter half of the tragedy is actually saying. For Young Hamlet grows up and grows dead in the same instant. 'The Court's a learning place' says Helena in *All's Well*, thinking of what may become of her love in the Court at Paris. Elsinore has been for Hamlet a 'learning place' too: more so, perhaps, than Wittenberg. But 'growing up', 'becoming mature', so easily taken for granted by us now as virtues, are in Elsinore's tragic Court as doubtful, as hard to sustain as Hamlet's brief ownership of his father's crown; and if achieved at all, these conditions bring death with them. It is in a grave, Ophelia's grave, that the Prince at last and for the first time identifies himself with his father, taking on his father's royal title: 'This is I, Hamlet the Dane.'

At the beginning of the fifth Act Hamlet returns to Elsinore after an absence that makes itself felt as marked. When he comes back to it again, his mysterious encounter with the pirates serves to embody time passing, and to make believable those subtle changes often visible in human beings only when they return after absence. Learning for Hamlet is a kind of departure, a going away from his origins; and he comes back different. The play even hints by the word 'naked', of course metaphorical, used in Hamlet's letter, and by his talk of his sea-gown, that he has left behind for ever his undergraduate, grieving, and courtly black clothes: the image that he makes is different, closer to that of Everyman. It is symptomatic, too, that we hear no more of the Ghost. His adventures on the high seas have been subject both to chance and to moral confusion—he has been saved only by pirates, 'thieves of mercy', and he has substituted for himself in the net of his fate two men once friends of his own. These bewilderments and defeats are a hard burden like the dead bodies of Rosencrantz and Guildenstern, joining the dead Polonius in the shadows of Hamlet's now extensively stained past. The Prince comes home, in short, through a graveyard, where he is just in time for Ophelia's obsequies.

The Court of Elsinore is what was once known as a 'man's world', one given up to the pursuit of power in a conventional system of rivalries. There is little place for women in such a world, and the women of this tragedy are markedly shadowy and faint. Gertrude can only signify her (doubtful) fidelity by moving from the side of her (politically) strong husband to that of her (politically) weak son; Ophelia is shattered by the conflict of her father, her brother, and her lover. Significantly, madness comes to this pathetic and evidently virginal young girl as the belief that she has been brutally seduced, almost raped. Her passivity is essential. And this is a point very interestingly underlined by the verbal by-play of the Clowns over the morality of her death—a passage unjustly neglected by commentators. The Grave-diggers' common sense, aspiring to legal expertise, tells them that Ophelia must have committed and been guilty of a willed action—she must have drowned herself. And yet, 'if the water come to him & drowne him; hee drownes not himselfe. Argall, hee that is not guilty of his owne death, shortens not his owne life.' She may even have 'drowned her selfe in her owne defence'.

In its context, and within the formal peculiarities of the whole play—shapely, logical, yet deeply undermining what we sometimes allow ourselves to hope about human freedom —this nonsense makes a curious and haunting sense. Ophelia is a shadow of Hamlet, a moon by his sun: 'cressant', as Laertes calls the Prince, her spirited yet defeated attempts to 'grow' in the presence of her officious father and aggressive brother only betray her, and condemn her Court existence to an end among 'Coronet weeds', the 'envious sliver', 'the weedy Trophies', 'the weeping Brooke', 'as one incapable of her owne distress'. But her very unfreedom transmutes real guilt in her death. Really, she 'drowned her selfe in her owne defence'—even, 'the water came to her & drowned her'. The preceding scenes of the fourth Act have distinctively entwined the morally dark with the natural—Claudius's corruption of her brother with the flowers the mad girl gives, the songs she sings, the Court's disturbed tenderness to her. That natural darkness, together with the silences of Hamlet's own failure, lie in the shadows of the Graveyard Scene as it begins, and even enter the Clown's songs about youthful love as 'solace', and about 'Age with his stealing steps'. And they convert the violence of young Ophelia's madness and death to something quieter, more profoundly natural, like the inevitable passing of time.

These things flow together as Hamlet comes back from the sea to hold the skull of the fool Yorick, a man who has in memory performed momentarily something like the role of Juliet's Nurse—the age of seven is surely used here, like Juliet's weaning and then her arrival at fourteen, as one of the great Elizabethan age-divides: it was the age for starting school or, in harsher social circumstances, for starting work.[17] Hamlet is suddenly in the presence of his lost childhood. This is one of the moments at which Shakespeare, writing within an Elizabethan culture that neither cared greatly for children nor took much interest in them (attitudes shared to some extent by the writer's own work), reveals startlingly how he has transformed a story of Crime and Punishment to a tragedy of experience, in which—to use Yeats's trenchant phrase—'The crime of being born | Blackens all our lot'.

When Hamlet thinks of Ophelia later in this scene, he moves into the past tense—'I loved Ophelia'—with the kind of clarity and simplicity he was far from finding as her lover; his at last loving glimpse of her pastness is like that strange, silent encounter recorded earlier through Ophelia's own memory. The two young people really only meet in their

[17] See, e.g., Philippe Ariès, *Centuries of Childhood*, tr. R. Baldick (1962), 66.

imaginations, even their fantasies (the Hamlet we see is superb, but he is not Ophelia's 'expectancie and Rose of the fair State, | The glasse of fashion and the mould of Forme': the delicious young girl's images are faded, stereotypical). Similarly, the earlier undergraduate was simply more at ease with members of his own sex, both adoring and frightened of women, seeing only his mother through them. In the past tense he is certain. But Hamlet can use the past tense because he now has a past tense—he has, as used to be said of women, 'a past'. Once human beings have a past felt as dark, as irrecoverable, and as their own, their life is beginning to be over. They are in any case no longer young.

For Hamlet, 'growing up' is also growing dead. He has reached thirty (the Gravedigger tells us), that ancient agreement as to the year of human maturity: but he will not live much longer. This arrival simultaneously both at full manhood and at death gives the play its mythical quality: it drafts out one of the great human rites of passage. But the play's progress is uniquely realistic as well as mythical. Hamlet's growing is given some of that haunted naturalness with which the water reaches Ophelia; merely waiting, watching, feeling, existing and always talking, Hamlet becomes something more than the busily conforming Laertes and the marching Fortinbras, who are no more than modes of behaviour. Supported as he is by the highly original and existential rhythms of the play, divergent, lingering, contradictory, and accidental, Hamlet's growing becomes a statement of being in itself, of human experience.

One of the characters in a novel by Ivy Compton-Burnett, *Elders and Betters*, says of the painful growing-up of the book's children, 'The process of getting used to the world seems to be too much for us'. The novel is one of those innumerable English fictions about which we can feel in passing that they perhaps might not have been written if Shakespeare's tragedy had not existed. At all events, that dry and trivial phrase 'getting used to the world' could be used to gloss the deep and subtle power with which the very end of *Hamlet* is handled. A completely new character arrives, called in the Folio 'young Osric'—a menacing young fop, and a King's Man in person—whose triviality is the keynote of all this last movement. Hamlet himself finds a new tone, adult and grim, light and disturbed, wearily impatient now to have things over and done with: 'I shall winne at the oddes: but thou wouldest not thinke how ill's all heere about my heart: but it is no matter.' His formal Court apology to his adversary, Laertes, brilliantly sustains the same tired and impenetrably public surface, his manner like the duel that follows honourable, empty, and—because empty—lethal. Young Hamlet's time is all but over. Its real end is soon marked by words from Horatio:

> Goodnight sweet Prince,
> And flights of Angels sing thee to thy rest.

Horatio's 'Goodnight' needs a word of comment it never seems to get. It might almost seem too sweet, too sentimental for the play. If it doesn't, this may be because of a shadowy irony the lines carry in themselves. Horatio, we feel, might almost be talking to a child—but perhaps he is talking to a child. Holding as they do that private sound of a nurse's good-night many times over delivered to well-born well-behaved Tudor children, the lines perhaps contain all the ironies of the play's own good-night to Hamlet's compromised youth, his freedom and his life.

And this possibility is conceivably strengthened by a curious fact in Shakespeare's own life. Biographers sometimes speculate on the relevance to the tragedy of the death, some four or five years before the play was first performed, of Shakespeare's only son, the eleven-year-old Hamnet. What doesn't however seem to be noticed is that in the very year of the boy's death Shakespeare's father—or the writer on his behalf—successfully applied for a coat of arms, thus ambitiously attaining the status of Gentleman. These two events perhaps became one in Shakespeare's mind, the seed from which his tragedy of a son began growing.

2
Othello: Mixing

OTHELLO was often known in its own time as *The Moor of Venice*, a title that interestingly parallels *The Prince of Denmark*. And the two plays, almost certainly written in sequence over no more than three or four years, the first of the seventeenth century, have concerns in common reflected by these titles. But what strikes the attention first is the contrast made by *Othello* to the earlier tragedy. Darkness and enigma give way in *Othello* to a hard, clear light. The drive of the single action permits the small group of characters to destroy each other with speed and economy. Magnificently written, afflicting yet in a sense external, *Othello* is what it is always said to be, Shakespeare's triumph of pure theatre among the tragedies.

'Theatre' is of course not the same as grand opera. It's sometimes possible to wonder whether our corporate image of the tragedy hasn't got fused with that of the stage's most spell-binding adaptation of it (rather as Shakespeare's *Lear* did with the Tate version in the eighteenth century): in short, with Verdi's *Otello*, first produced just a hundred years ago. *Otello* is potent because music does work on us more directly than language, and because Verdi's 'line' is—or is often said to be—less confused and cluttered than Shakespeare's. What isn't always added is that Verdi and his librettist achieved this clarity by dropping Shakespeare's first Act and radically simplifying his last scene, so that the opera begins with the arrival in Cyprus and ends with the murder of Desdemona. By doing so they concentrated on Othello's love and marriage, shaping the work so that it becomes that formal construct of feeling which music always provides. But to assume that literature too consists only of forms of feeling may be mistaken. If Shakespeare's 'line' is different from Verdi's, this is because what confuses and clutters it is meaning—that meaning which even the greatest opera can manage without.

Othello is now almost universally called a Love Tragedy, even a Domestic Tragedy. The story of sexual intrigue and jealousy which Shakespeare took over from his main source, a novella by Cinthio published in Venice the year after the dramatist was born, gives the tragedy a substance which makes these terms not meaningless. But to read Cinthio's harsh tale, as to attend Verdi's grand opera, is to know that 'Love' may mean many different things: a fact important in life and literature alike. A category such as 'Love Tragedy' may offer a false authority, concealing its own tendentiousness, seeming to derive from some long tradition yet in fact (like almost all our common preconceptions about Shakespeare and his work) hardly dating back much further than Bradley and the first decade of the present century. Bradley was worried by a lack of 'universality' in *Othello*, and many critics have followed him in finding a certain meaninglessness in the tragedy. But Bradley's 'universality' is the product of Bradley's universe. The lack that he recorded was of a metaphysical absolute; and the means by which he attempted to replace it was by stress on that heightened morality of emotion which is often a staple of Victorian religious feeling ('Passion spins the plot'). The critic re-invented *Othello* as a Love Tragedy, a work curiously like *Otello*—composed less than twenty years earlier—but without the music: a drama about a passionate Moor who terribly murders his wife out of jealousy, but whose action can be considered heroic rather than villainous, indeed somewhat sublime, because of the quantity of intense feeling with which the act is endowed.

In the eighty years since Bradley's very distinguished essay on the play, many able critics not consciously close to him in approach have none the less similarly accepted an interpretation of its action which is Verdi-esque. The tragedy is seen without its first act, where the Moor stands free and his world is built up round him; even more, the vital and very curious last scene of the play is all but ignored. But this tacit plot-summary of the passionate jealous Moor neglects what, in my view, proves to be the tragedy's true crisis: Othello's discovery that he has murdered his innocent wife essentially because Iago (his Ensign or Sergeant-Major) told him to.

> I am not valiant neither:
> But every Punie whipster gets my Sword.
> But why should Honor out-live Honesty?
> Let it go all.[1]

'I am not valiant neither': the remark, with all its human truth, its tired petulance, its sheer comic hopeless irritability, shows Othello more honest than some of his critics. He understands, that is to say, the nature of the plot which here accomplishes itself.

The conventional 'plot' of *Othello* is after all strangely anti-feminist, even when asseverated by women; it sentimentalizes the play into '*Otello e Desdemona*, a Love Story', and yet it leaves the woman out—it leaves out of account Desdemona's implacable virtue of chastity. For *Othello* is not a play about a man who murders his wife; it is about a man who murders his innocent wife; it is about a man who murders his wife by mistake. Only this last phrase gives due value to the salient innocence of Othello's wife, and to his discovery of that innocence in the last act—a discovery which entails knowledge of his own terrible folly. To imagine a play about a Moor who murders his wife in jealous passion is to postulate a work that Shakespeare could have written better by choosing a plot in which a wife actually did act unfaithfully. This is what he came close to doing in both *Troilus* and the *Sonnets*, two works in which no problem of credibility (no 'double time scheme' or other such expression of incredulity) distracts the auditor from the central sufferings of love and jealousy. Cressida and the Dark Lady are infidelity *per se*, constructs from a plot that makes them what they are. But if the plot of *Othello* really centres on jealousy, then Shakespeare wrote it badly; for a fiercely faithful wife can do nothing but distract from the seriousness, the authenticity of her husband's condition.

[1] Shakespeare quotations are from *The Norton Facsimile*.

It is not hard to see something almost equally problematical in a play about a man who murders his wife by mistake. A snigger is so deeply latent in the subject that one can see very precisely how and why *Othello* upset, and will always upset, late-Victorian sensibilities that look to tragedy for heightened dignity, for beautification of life. But an element of farce is— as the Restoration critic Rymer realized, however insensitively —congenital to the play; and it was this which, to my mind, provoked at least a part of the intense pain Samuel Johnson felt (he called the last scene 'dreadful', 'not to be endured') as an Augustan always awake to social dignity and its loss. It is an important fact that all Shakespeare's other plays that deal in mistaken jealousy (*Much Ado, The Merry Wives, Cymbeline, The Winter's Tale*) are comedies or tragicomedies, all plays that give us formal expectation that something will distract us from the final centrality of the jealous hero: a saving distraction which secures him in the end, even if by virtue of diminishing him. In terms of plot structure, *Othello* has a plot which—as critics have noted—shows affinities with earlier comedies rather than with tragedies and histories;[2] and in its substance remains a work which the dramatist conducts, with the greatest artistic brilliance and verve, along the very edge of comedy. Yet even so, and with all its brilliance, it seems to me probable that textual differences between Quarto and Folio versions of the play may be explained by Shakespeare's attempting to correct some of the problems entailed by this leaning towards comedy—this inevitable diminishment of its hero. What might be called the indignifying capacity of the quasi-comic villain, Iago, strong in the essential superiority given him by the plot-situation, had gone too far: the dramatist therefore worked, through for instance the addition of the great 'Pontic sea' passage to Othello's role in III. iii, to increase and deepen his hero's already formidable dignity. The extension of Desdemona's song in IV. iii may be a similar heightening of her dignity by her power to move us.

[2] See, e.g., Emrys Jones, *Scenic Form in Shakespeare* (Oxford, 1971), 119–28.

Pervasive through the Renaissance is the assumption most familiar to us from the comedies of the Restoration period: that for a man to be cuckolded is a subject fit for ripe mirth. Shakespeare shows few signs, even in *Othello*, of sharing that joke. The 'comic' aspect of the drama therefore needs more careful explanation than the belief that in Shakespeare it somehow goes naturally along with jealousy. Moreover, whether or not the Moor *is* jealous, a topic much debated since it was first initiated by Emilia and Desdemona, is an issue requiring similar care. We can take out of a work only what its inclusions and exclusions have made available to us; the question of Othello's jealousy is perhaps unanswerable because immaterial. In a play that is not precisely a Love Tragedy, Othello is not conclusively jealous, just as Iago is rarely very funny. All these possibilities are latent in Shakespeare's story as he found it in his source. What he made of that source is a very different matter.

In 1565 Geraldi Cinthio put together a large collection of novellas mainly concerned with the social manifestations of love. One, beginning 'There was in Venice a Moor . . .', tells of a jealous young Ensign who can't get the woman he wants, the wife of his General, the Moor, and who therefore drives the Moor into murdering her, in a mad jealousy and helped by the Ensign. The story runs briskly and compellingly from an incisive beginning describing the couple's established and happy married life in Venice, moves swiftly after a few paragraphs to Cyprus, and at length achieves a brutal if long-drawn-out close. Its early stress on wedded happiness only intensifies a quality of painful randomness in the story as a whole (a randomness which survives a professed moral purpose to discredit mixed marriages).

But a certain randomness could be said to be inherent in the very genre of the novella, a form which anticipates the newspaper as much as it does the novel (it is not irrelevant that the stories of the *Decameron* are prefaced by a description of the arrival of the plague in Florence). Shakespeare, whose responsiveness to literary forms was acute, took every other tragedy he wrote, even including *Romeo and Juliet*, from historical materials, or from materials that passed in the Renaissance for historical. As a result, his tragedies characteristically play off a context of contemporary allusion against an ancient and true story, a practice that always deepens and vitalizes their sense of reality. This juxtaposition of the historic and what we make of it in the present is particularly ironic and touching in *Hamlet*, where sons are continually tolled back into the past of their fathers by the bell of their own present, striking 'One!'.

Othello is Shakespeare's only tragedy set entirely in the present. The Moor himself has a different retrospective dimension, something like a rich landscape of memory, but it is always romantic, a matter of 'antres vast and deserts idle', and in the course of the play grows—as in the history of the handkerchief—doubtful or ironic. This peculiar presentness derives from Shakespeare's choice of a novella rather than of a history as a source for his new tragedy; and the choice can scarcely have been accidental or unconscious. But Cinthio's story gave more than its form to the dramatist. The novella's Moor was an established Venetian. It is possible that this story, itself published in Venice and concerning a Venetian, revived in Shakespeare's memory a great complex of ideas and suggestions which finally come together in his Venetian tragedy. They help to suggest why Shakespeare chose this random story, and why he changed it so much, and yet why the one thing in a sense he didn't change was its randomness. The Venice of *Othello* is bare in the kind of topographical details the dramatist had made good use of in *The Merchant of Venice*. And yet it is the same place—an interior place, a kind of psychic geography. Critics often notice how problematic the young lovers of *The Merchant* are, even Portia, Shakespeare's first great lady of comedy, who with all her grace and resourcefulness can only trick Shylock. A certain meanness touches all of them, which can become a surprising obscenity in the backchat; and perhaps the most attractive of the lovers, Antonio, is hardened by habitual melancholy. Belmont is moonlit, but dark shadows fall across the market-place down below. Money is the one thing everyone has in common.

The Merchant of Venice is in short a very original legend of the rich, for whom Love and Money are terms of each other; and in this legendary world love cannot be gained without the simultaneous losing of a gold ring of fidelity (perhaps the one thing about Venice most of Shakespeare's audience would have known was that the great trade city had wedded herself in love to the ever-changing and destructive sea). *The Merchant of Venice* comes near to inventing a milieu hardly met again in literature before the modern period, in novels like Henry James's *The Wings of the Dove* and *The Ivory Tower*, Ford's *The Good Soldier*, Fitzgerald's *The Great Gatsby*, even Amis's *I Want It Now*: this is a romance of millionaires, a world of the beautiful and—if not damned—only somewhat ambiguously saved.

When, in the second Act of *Othello*, Iago is entrapping with drink his successful rival Cassio, who has got the Lieutenantship that Iago says he hoped for, the romantic, honourable, and probably rather rich young officer seizes the chance between hiccups for a little moral reflection:

CAS. For mine owne part, no offence to the Generall, nor any man of qualitie: I hope to be saved.
IAGO. And so do I too Lieutenant.
CASSIO. I: (but by your leave) not before me. The Lieutenant is to be saved before the Ancient. Let's have no more of this.

(II. iii. 111–16)

By birth a Florentine, Cassio is by blood a Venetian. After one glass his theology deliquesces into snobbery, just as his new rank disappears in ten aggressive minutes more: value is here as baseless as gains on the market may be. Though the Lieutenant hero-worships Desdemona, his General's wife, his actual mistress is a prostitute (a character added to the action by Shakespeare); and his use of Desdemona to get his credit back with Othello is not without calculation, even if by Iago's prompting. And Cassio is, as it happens, first introduced to us in the play—even if, again, only by the viciously untrustworthy Iago—as an 'Arithmatician', a 'Counter-caster'.

A little earlier again in this same scene, the first in the play, Roderigo—Iago's own pocket Cassio or Lieutenant, a foolish gentleman who is another invention of Shakespeare's—opens the tragedy with three lines unforgettably if obscurely defining Venice for us. It is a place where—as in *The Merchant*—Love and Money are one and the same thing: 'I take it much unkindly | That thou (*Iago*) who hast had my purse, | As if the strings were thine, should'st know of this.' Roderigo accuses Iago of not behaving as lovingly or loyally as his use of Roderigo's funds should have guaranteed, a betrayal proved by his being so *au fait* with someone else's affairs. And, because both 'purse' and 'knowledge' have faint carnal under-meanings in Elizabethan English, love and money, public affairs and private, the sexual and the mercantile all fuse in the play's first lines, a world of hints gathered up into the control of Iago, who knows all the uses of love: 'I follow him to serve my turn upon him.'

I am suggesting that Cinthio's novella was chosen by Shakespeare because its Venice, and its Moor, together with the Italianate context of the form itself, so acted on the poet as to draw towards daylight and multiply rich associations already potential in his earlier drama of Venice. These associations were in part inherent in the whole fabric of Elizabethan Petrarchan love-writing; the Sonnet, for instance, is a courtly form, its jewel-studded content reflecting its concern with the uses of love. Moreover, Petrarchanism was not only an economic but a political medium in the sixteenth century, just as the actual Moors of the poet's time had a political context real and troubled enough: at the time of the writing of *Othello* Shakespeare's London held a considerable number of impoverished Moorish refugees from Spain, where the Spanish Moors, no longer noble or wealthy but citizens for a millennium, were within a half-dozen years of their final expulsion. That Shakespeare's Othello is unlike Cinthio's Moor in being a 'stranger', an alien and a newcomer, is in my view much more important to the play than the more modern issue of his 'colour', which is introduced by the hostile and trivializing Iago and spreads from him like an infection.[3]

If Shakespeare's contemporary Moor was not without politics, Venice had for Elizabethans its double image. For Shakespeare's age it was the greatest of trading cities. But its wealth had had political uses: and Venice was renowned as the one great Republic of Renaissance Europe. The Republic is to the Court as the novella is to the history. The Republic lacks any metaphysic of the political, just as the novella lacks the dimension of human pastness. On his first appearance, already made defensive by Iago, the Moor speaks of royal blood: 'I fetch my life and being | From Men of Royal Seige.' But in Venice this is a vain boast. The claim to blood royal may dignify Othello in his own eyes but gives him no real power. This hint of exile, of isolation and perhaps of danger, is intensified by the posting to Cyprus, an ultra-Venice or outpost of Empire. If Venice has no royal house it possesses at least its own grandees, and the Duke is just and rational as well as powerful. In Cyprus Othello governs by a dignity unsupported from without and threatened by its own shadows within.

I proposed earlier that Hamlet might be thought of as a man growing. Though past his youth, perhaps past his prime, Othello has his own emergence. In the first Act of the play, called before the Senators to explain his elopement with the daughter of one of their number, the Moor pleads lack of verbal sophistication, having served with the Army since he was seven years of age 'Till now, some nine Moones wasted'. The phrase is lightly dropped. Yet it leaves behind, it may be, some faint suggestion that though Othello's childhood is far in the past (at seven one leaves home for school or work),[4] none the less there are conceptions and births into this world for adults too. For that 'nine Moones' or months may do here what it does dangerously at the start of *The Winter's Tale*, and

---

[3] See Chap. 9 below.   [4] See Chap. 1, n. 17 above.

measure a coming-to-birth. Something is beginning. Othello's emergence has to do with love—with his great feeling for Desdemona, who alone bonds to Venice his 'unhoused free condition'. But it is a love which he has to defend to a Senate, as he is doing now, against the angry protests of Brabantio (the sudden news of whose death will near the end of the tragedy throw a long shadow back across Othello's elopement). And that elopement itself is communicated to us, in so far as we understand it at all, through the contorted, ambitious, obscene gossip of Iago and Roderigo, and of both with the 'robbed' Brabantio, before the Moor himself is seen in the play.

Love for Othello is a beginning, an arrival—but an arrival always at Venice. I have argued that Shakespeare chose his source—whose subtitle might almost have been 'Death in Venice'—not in spite of but because of a certain meaninglessness inherent within it. On its thin fabric the dramatist built up a tragic society that coheres with itself: a society whose tragedy is that it is as 'thin' as the source. The genuinely high-souled Moorish stranger absorbed into this society in the end loves not his wife but the world of surfaces that breu her; he is Iago's ally not the lover of Desdemona, who dies in her wedding-sheets. Gradually isolated from all who love him and from all he loves, Othello moves into a world as social as possible while being as little truly sociable: a condition more deeply destructive of love than the sexual jealousy which Othello nominally suffers from, or is persuaded by Iago to cultivate in the name of his social honour.

The novella of sexual love and jealousy was through each of its terms, both in form and in substance, committed to a world of pure and immediate *use*, a living in the present. In Shakespeare's play that presentness has taken on tragic dimensions. In his grief over the wife he has murdered Othello assures us

had she bin true,
If Heaven would make me such another world,

Of one entyre and perfect Chrysolite,
I'ld not have sold her for it—

(v. ii. 143–6)

but in fact he has sold himself, if not his wife, for just such a world, as wonderful and as small as a world-sized bauble; and his final dignity-retrieving jewelled boast to the Senate makes it plain that he knows, if nothing else, what a fool he has been. But the dramatist is both humanist and comedian enough to count 'folly' intrinsic to real human life. Shakespeare's ageing and alien but magnificently dignified hero, in the course of his story bruised, bewildered, violent, bad, hideously mistaken and in his undeception hideously deprived of dignity, could perhaps be described as for the first time in his life actually living. The story of love's confinement in a rich Republic in short offered Shakespeare the perfect medium for a definition of human experience itself, solid, absolute, and self-destructive.

This double process, at once of advance into love and life and of destructive corruption by society, I have tried to gloss by the modern colloquial term, 'Mixing': a word which oddly enough in Elizabethan English covered all the senses I have hoped to use it in here, from sexual intercourse to social congregation. The play itself 'mixes', its shimmering time-scheme making it seem at one moment as if the action spanned (as with Cinthio's couple) long years of steady comfortable social existence as married people, and the next as if the play lasted a time not only too short for adultery to have taken place, but as if the whole course of it comprised the consummation of Othello's union with Desdemona, from the unseen elopement to the final murderous marriage-bed ('Tonight | Lay on my bed my wedding sheets, remember'). So, Iago playing with the words 'serve' and 'service' casts strange shadows on Othello's interrupted and delayed wedding-night: shadows that darken until he 'dies upon a kiss'. So, too, it might be said that with the elopement, that impulse both into and away from Venice, begins—despite his denial of desire—Othello's learning of the corrupting lesson what it is in a worldly sense 'to have and to hold'. Having and holding is an essentially Venetian affair; and jealousy too is a terrible possessiveness or ownership in love, a robbed collector's pain. If jealousy is a passion, in Shakespeare's eyes it is an extremely social passion.

Society is a place in which we can only say what we know everyone else will understand or tolerate. Its most characteristic style is therefore small talk. But this is a language which can create peculiar difficulties in the expression of private and personal depths, or of absolute moral judgements. In this grasp of social speech, as in a startling number of social factors, *Othello* seems a prevision of the English culture eighty or a hundred years after it. Throughout the Restoration and Augustan periods the heroic was brought to terms with small talk, like Gulliver among the Lilliputians, and the result was a plain speech in a resonance of ironies. In terms of social behaviour, the good man became the 'honest' man ('honnête homme'), of which our own best parallel is the nice man. One of Iago's more gentlemanly descendants, or literary parallels, is Jane Austen's Mr Elliott, whose viciousness is at last perceived by Anne from the fact that everybody likes him—*too many* people like him.

But Iago is, in fact, more original, even more 'modern' than this. His brilliance as a character is his blankness, his commonplaceness, in the end his uninterestingness: his complete adjustment to the Venice he remakes in his own image. Despite all their jokey intimacy, his soliloquies are fake—though we may sharply recognize him, we never see into Iago, who is a construct of social attitudes and social appetites. From this follows the fact of the invisibility of his villainy to the other characters. It is not that they are particularly stupid as individuals (though they are not particularly intelligent, either); but that Iago wears the 'invisibility' of the sheerly social unit—as we call a social lie, 'politeness'. Emilia's strange, striking, and reiterated 'My husband?' in the fifth Act makes it seem almost that she has genuinely never seen Iago before; she has only accepted the 'anybody' in the role, the driving male non-person who inhabits her bed. And Iago has a public persona which matches this private one, one which shows Shakespeare's power of social imagination at its most extreme. Iago plainly sees Othello's relation to himself as that of Master and Man, a relation he perverts and inverts. As a result he exists as a vortex of continual power-energies. Dramaturgically perhaps descended from Shakespeare's Rumour, and related to his Lucio too, Iago is the contemporary crowd packed into a single consciousness: he is the voice of the Mob, and trouble flows mindlessly from him,

> As when (by Night and Negligence) the Fire
> Is spied in populus Citties.

Iago is, in short, any member of the Venetian back-street crowd, but he is that in a crowd which pushes to the front. It says something about Iago's extraordinary originality as a character that the critical study which throws most light on him is possibly Canetti's psycho-philosophical (and very twentieth-century) *Crowds and Power*.

The 'honest' man is one who admits that in society only certain things can be said and done, and they can be said and done only in certain ways. But Iago exists before such concepts were usable because systematically formulated. And he exists as the servant of a man who declares his own alienness to any such society: 'Rude am I in my speech'— Othello's token is not Honesty but Sincerity or Authenticity, the belief that anything can be felt and that anything that can be felt should be said. If Iago lies, Othello is perhaps held by fallacy. It is very terrible that when Othello feels himself to be suffering most, what his mouth says is 'It is not words that shakes me thus, (pish) Noses, Eares, and Lippes: is't possible. Confesse? Handkerchiefe? O divell.' Face to face with simple Othello, nothing is more dishonest than the pseudo-civilized honesty of honest Iago. But then perhaps nothing is more foolishly arrogant than Othello's falling in love in Venice— even falling in love *with* Venice—without once stopping and looking fully at Venetian Iago, who is his own dark shadow, the Master's Man.

Othello happens to be one of the few plays which in its Folio version is followed by a Cast-List, possibly not by Shakespeare himself; and this, while casually and informatively referring to Othello as '*The Moor*', and to Cassio as '*An Honourable Lieutenant*', names Iago '*A Villain*': an odd usage, perhaps not dramaturgical, but the common colloquialism of the time, as if to remind us that readers and audiences are a society too, and that like society we lightly recognize what Iago is, but have no intention of doing anything but letting the play go on. Society is where Honest Iago lives. His vitality is, moreover, enmeshed with its continuity, because society, unlike individual life, doesn't stop. *Othello* is Shakespeare's only tragedy (but for the comparable *Titus Andronicus*) in which the villain survives at the end, and though Iago may be mute and awaiting horrible torture, we somehow don't expect his death. Society is, after all, a place for survivors. In this way as in many others, *Othello* is a comitragedy, accepting the survival of the unfittest.

This social aspect of the tragedy explains features of the drama often found difficult. A scene almost universally disliked opens Act II, where the action moves to Cypru Verdi's *Otello* starts here, beginning with an orchestral thunderclap as a sail is sighted from the quayside, victory over the Turks declared and exulted in, and the arrival of the Moor himself effected. Othello's arrival has a wonderful tenderness in the play too:

> OTH. O, my faire Warriour.
> DES. My deere Othello.

But in Shakespeare the lovers' meeting is prefaced by the tedious and trivial backchat of Iago, Emilia and Desdemona, the chronicling of 'Small Beere'. Commentators question the purpose of this sequence. What we have here is surely a re-establishment in the colony of the system of Venetian Small Talk. Desdemona is a good and loving woman (that her marriage to Othello is to frighten her into telling lies will be a part of *her* tragedy). But she is also a Venetian Senator's daughter: she is 'in society'; and like the virtuous social heroines of comedy, Desdemona politely plays her part and laughs at Iago's low jokes.[5] As a whole, the scene has an effect impossible for even the greatest nineteenth-century opera. In the concatenation of Iago's obscene asides and the General's sudden arrival, 'Cluster-pipes' next to the bronze notes of the Moor's 'Trumpet', there is a shock of the real, a disturbing actuality in the establishment of the mood of love in society. Love is a magnificent secret that Othello and Desdemona tell each other on the quayside. But there is also fear in the secret, and Othello feels it.

At this point a reservation is necessary. As I hoped to suggest by mentioning the Cast-List's jocular 'Iago. *A Villain*', social converse is by definition morally neutral. Iago's 'Small Beere' is a very different thing from society's small talk. The breath of fear, or danger, that we feel on the quayside comes at least in part from the fact that Iago is 'nothing, if not Criticall'. His contemptuous degradation of women's domestic function (he sees no other) is far from the regular talk of the weather in, say, Jane Austen's Highbury: where the tedium of Miss Bates might be said to derive from the incessancy of her good nature.

A brilliant tradition of high social comedy runs through English literature, from Chaucer to Firbank and Barbara Pym, taking in on its way, along with *Emma*, such things as *The Way of the World* and *The Importance of Being Earnest*. *Othello* is not much like them. The distinction is that, within its world of surfaces, the tragedy finds horror—we could almost go on to say that it understands void. Some of the post-Bradleian criticism that lays weight on the 'Nobility' of Shakespeare's Moor at the same time insists also on the 'beauty' of the work itself. Works of art are of course beautiful, and *Othello* is no exception; and its Venice may be called so, even if some of the play's most splendid evocations and images are (other critics, the 'Ignobilists', would argue) hard and sensual in their beauty: even, I would add, elusively

[5] A close parallel is Helena's conversation with Parolles in I. i of *All's Well*.

mercenary—Othello's ravishing world of Chrysolite is a case in point. But it is surely necessary to say that *Othello* is beautiful *and* very ugly; that the play features passages almost more startlingly, needlingly obscene—with something farcically dark in them as well—than almost anything else that Shakespeare ever wrote. If this horror is to be called 'social', careful definition is required.

Primary here is III. iii, the great 'Temptation Scene', with Iago's invention—before a fascinated Moor, muttering rightly, 'O monstrous! monstrous!'—of Cassio's dream of lovemaking with a Desdemona who turns into Iago himself. Othello's first two soliloquies in the play (at I. iii: 'Most Potent, Grave, and Reverend Signiors', and 'Her Father lov'd me, oft invited me') give him a past that is like a dream, an imaginative possession of romance and mutual hospitality, mutual love. Iago 'dreams' too—or rather he unprovably, inarguably conjectures a past that contains Cassio's conjectured dream of love and lechery: a dream that turns into Iago. The passage, echoed painfully at IV. i—'to be naked with her Freind in bed, | An houre or more, not meaning any harme?'—is 'Mixing' with a vengeance. The horror here does not lie in Iago's seamily polymorphous sexuality, of course self-centred. It does not even lie in the whirlpool of fantasy Iago's presentness opens out to reveal. It consists rather in the fact that Othello listens, and having listened, joins. The strikingly fastidious, almost perturbingly celibate warrior who, perhaps in an attempt to protect his privacy, denies desire before the Senate, is himself 'Mixing' with Iago, drawn into his detestable dreams. The peculiar edge on the situation is given by what is social in it. In Othello's brave evocation of 'the general Campe' enjoying poor virtuous Desdemona's body there is a sudden terrible laughter which hovers pervasively throughout, and which is a measure of the way these scenes balance depths against social detachment. The Moor is so conscientiously trying to 'be a man', trying to be 'worldly' in Iago's suffocatingly small and vicious social definition of these terms: a definition that explains why all the centre of the tragedy turns on 'looking on', on voyeurism, on proof—in a word, on the losing and finding of a mere handkerchief.

In the course of the play, Iago corrupts Othello. But the process is two-way: Othello transforms what would be merely a trivial cynicism in Iago into what is, in his own experience of it, morally horrifying. The experience is actual and is Othello's. Throughout the play, that Desdemona is chaste is as evident as it is unprovable; as unprovable as the dramatist has carefully made everything Iago alleges here and everywhere. Hearsay, as English law declares, is no evidence. In so far as we are social, what matters is what happens before our eyes; and what we see is Othello, assenting that Venice should become a structure of dark and stolen secrets: an assent that violates everything else that he has believed it to be, and himself as a part of it. This act, of making the social susceptible of profundities of experience, is Shakespeare's great original achievement in the tragedy. The story of sexual intrigue enlarges to hold 'Crowds and Power'.

I have been arguing that in *Othello*, a 'Love Story' comes to hold breadths and depths of social experience. But 'society' too is interpreted in the play. Shakespeare's Venice, great trade city and Republic, requires the heroic Moor—hence his final laconic 'I have done the State some service, and they know it'. But Othello would not be Othello without that royal blood equally reticently mentioned in the Moor's first full speech, a claim laid down like some counter at a roulette table: 'I fetch my life and being | From Men of Royal Seige'. Othello's 'royalty' is much more important than we might assume. It is the clue which connects this tragedy with the other three great tragedies, where Hamlet is Prince, Lear King, and Macbeth one who hungers for kingship to the point of murdering for it. I want to pause very briefly to consider what Shakespeare makes of royalty or kingship, not merely because it affects manifestly the other three plays, but because this concern goes deep into *Othello* also.

These tragedies' involvement with kingship would now be glossed as simply political. But Shakespeare's handling of the theme is large in all its relations with common human experience. Part of the intense interest of *Hamlet*, and of its initiating force, is the way in which the dramatist has in it translated 'royalty' into the issue of essential human dignity. An ancient classical teaching concerning power and status in life was widely diffused in the Renaissance period; it would have interested a dramatist most in its insistence that all worldly fortunes are no more than roles, to the human actors cast in them at their birth. Prince and pauper alike act out parts both unchosen and subject to change: a Stoic belief that helps to release 'royalty' from politics into metaphor—as it does in *Hamlet*, a play peopled with Kings and actors, its politics one of change and uncertainty. Hamlet's own claim to the throne, his royalty in a political sense, is as unclear throughout as is the question of who his true King-Father is. Royalty comes to him as the necessity to live through the *in*dignity of sonship, the great human abeyance before the 'whips and scorns of time' as summarized in that elemental inheritance of family, social milieu, historical context. The command from one King-Father to kill the other holds the Prince within the indignity of being a mere mechanism of revenge, which in this play becomes a formal structuring image of the historical trap in its closure on the individual. For Hamlet, not to obey is the only resource of a royal dignity: to watch, to wait, to be 'outside'. Trapped between dead and living Kings, held by his love as by his hatred, Hamlet finds in the *not* doing his one refuge of spirit. His tragedy is that this refusal gives him the time to find his insubordinate dignity corrupted into the image of that Power that constrains him. Hamlet 'would have proved most royal', but dies merely the King.

In *Othello* there can be traced the lineaments of the same concern, the more fascinating in the absence of a Court as support and framework. A figure of dazzling and true dignity ('Keepe up your bright swords, for the dew will rust them'), feared and resented by Iago and Roderigo, respected by the Senators, hated by Brabantio, and deeply loved by

Desdemona, Othello steps out on to a stage whose drama will be the tragic losing of that dignity:

> know *Iago*,
> But that I love the gentle *Desdemona*,
> I would not my unhoused free condition
> Put into Circumscription, and Confine,
> For the Seas worth.
>
> (I. ii. 24–8)

The dramatist begins his story (unlike his source) with the elopement. And this original initiating gesture, as violent as it is romantic, is one on which the play is circumstantially tacit. We can only explain it by saying that by it Othello has protected his dignity—which can only have feared refusal—in a fashion dangerously close to an act of sheer power. Elopement, which parallels perhaps in a far more innocent form that initiating action of the previous tragedy, the murder of King Hamlet, gives to Othello's marriage both its independent autonomy and its spoiled social or worldly image of love as theft or robbery—love as a giving that is first a taking: here first Othello is seen as 'purchasing experience'. The elopement bespeaks the ambiguity with which the Moor's dignity domiciles itself in Venice, which is rich and yet potentially corrupt enough to convert that very dignity to indignity; and it makes us know—at least through Brabantio's death, if by nothing else—that if Othello is 'royal' then somewhere in his power there is a ghost of tyranny.

Little of this weighs on us at the beginning, when we first hear of the elopement, because any blame or shame the event could have is shunted off on to the figure of Iago: who is as simply a bad man, as the world goes, as Othello is—as the world goes—a good. But, for all that, Iago is from the beginning something like a negative or dark shadow of Othello: different, but not unconnected. His condition in the first scene has to be described as an enraged loss of power experienced as a loss of dignity: than which there could be no neater reversal of the elopement, which is a safeguarding of dignity that descends helplessly to a seizing of power. We are the more deeply if subtly troubled by the ambiguities of the Moor's relation to Venice because we incessantly see it reflected through Iago, the friend and servant who is also—being born Venetian—the social agent of this foreign and distinguished General. Indeed at the beginning we see Othello absolutely through Iago's sense of him, the two of them 'Master and Man' together—just as Iago carries Roderigo around with him as his own 'Man' (like the parcelled-up fly the spider stores in its web). The Othello we see through this refraction is involved with Iago in one action that has two mutually reversed faces, the two together speaking of the interconnection within Venetian society of the power of dignity (which Othello loves) and the dignity of power (which Iago lusts after).

Iago sees in Othello, as he would in any other Master, a greater dignity which afflicts him as the diminution of his own effective power ('Were I the Moore, I would not be Iago'): a diminution that must therefore be effectively reversed. This is pin-pointed by the fact that Iago has, by his own account (only as valid as anything else he says in the play), just been passed over for promotion. Iago feels this lack of dignity as lack of power; hence his ironic, even sinisterly 'poetic' relish in regaining power by the very act of telling of humiliation, which is a kind of verbal mastery. He sees power as to be gained, in a horrible mockery of the elopement, by 'violating' Othello's image and degrading his dignity: hence the fact that throughout this first scene Iago's speeches hold dark undertones of a parodying sexual relationship. At least two distinguished actors, Olivier and MacLiammóir, have—presumably for this reason—read Iago as concealing a homosexual passion for Othello: mistakenly, I think, on larger literary grounds. It gives Iago an interest, an emotional inwardness which humanly he never possesses. But the conjecture sees something that is really there. Savage and sinister puns like the sexual ambiguities in 'I follow him to serve my turn upon him' effect from the beginning a corruption that haunts the public face of love.

To argue that *Othello* is no simple Love Story, nor (even more) a Domestic Tragedy, is not to deny that the work carries a considerable charge of sexuality. This sexual charge originates from Iago, and is a function of his obsession with power and potency. I spoke earlier of his crowd-like amorphousness, his social invisibility, his absolute lack of depth. But he does possess a faceless sexiness. Lacking the dignity of the true individual, Iago's only function is to 'do'—though because his power is limited by subservience, most often the 'doing' is by proxy: hence the rank sexual fantasy and the congenital aptitude for jealousy. 'Do', like 'have' and 'know', is one of those monosyllables which retained for Shakespeare's audience a cant sexual force underlying other and more public usages. That these words recur with frequency in the play is important. Iago converts an action engaged with large issues to the level of 'do', 'have', and 'know'. Othello follows Iago and accepts that human dignity is the power to do, to have, and to know; and accepting this, he is lost. The 'Venetian' postures are on him nothing but shame and indignity.

To trace this process it is only necessary to recall the play: the whole of whose action is a steady and desperate gathering of power into Othello's hands. The key 'Temptation Scene', III. iii, knots all the earlier threads. The Venetians' staying on in Cyprus without the rationalization of military activity generates an energy of anxiety exacerbated by Cassio's loss of control—an action which implies a threat to Othello's own control over his men. Desdemona's own headstrong if generous support of Cassio extends that male threat to the General's dignity by something new: the much larger if sweeter challenge given to male superiority and status by the undermining, unhierarchical directness of the female, unimpressed by male ranking. In her the wife assumes the formidable upsettingness of the mother or nurse: ''Tis as I should entreate you weare your Gloves'. The male dignity of heroic solitude can perhaps withstand even less well than masculine rivalries a female insistence on the wearing of gloves. In effect Desdemona brings into play immediately *before* Iago starts his work in the Temptation Scene what one might call the terrible equality of love: a human response neither 'above' nor 'below' but assuredly equal and other, companionable. And it is 'terrible', in this context, because the strategies of power are already in the air: there is between the already faintly irritated husband and wife a good-humoured but actual tussle for the lead. Desdemona's innocently brutal 'Stand so mam'ring on' is answered by Othello's gently placing 'Excellent wretch': which, though often averred by critics to be harmless, would not have been used by (say) Essex to his Queen without repercussions.

It is in this context that Iago acts. He is all geniality, all submission: he openly eschews the 'terrible' in human relationships, the unpleasant, the grating, the appearance of quarrel. His assumption of the stance of the caring friend is such as to evoke a prelapsarian cameraderie of 'all boys together', safe from the invasive onslaughts of women. But under this surface of joviality his real address is half obsequious, half dominating. He initiates Othello into games of Master and Man, into pleasures half servile, half sadistic. He teases the Moor by threatening to withhold knowledge which at the same time promises mastery: 'Thinke, my Lord?'; 'I am not bound to all that Slaves are free to'; 'When most she seem'd to shake, and feare your lookes, | She lov'd them most.'

With these lures, Othello accepts Iago's terms, the lowest terms of the Venetian market-place, whereby power, to exist, must be seen to exist—the Renaissance equivalent of conspicuous expenditure. Othello takes up the gauntlet thrown down by Iago and accepts his challenge of power: he will prove himself master. He can be seen at last, that is to say, exchanging dignity for power. For true human and moral dignity could have been preserved; all that was needed was a formal refusal to discuss his wife with any other person in this manner. But this would have entailed (paradoxically) loss of face. Othello loses his power over himself by needing to prove his power over Iago; he loses his dignity by saving his face. After his hideously naked 'fit' in the fourth Act, a moment that brutally defines Iago's hunger for power, Othello rises wearing his new social mask: he will turn on the world the successful dangerous face of the Jealous Man, the Man of Honour. It is a part, in the event, which Othello saves himself by acting very badly indeed, in a grotesque Restoration rhetoric never properly commented on by critics (''Tis the plague to Great-ones . . . 'Tis destiny unshunnable'). Power and folly together shadow the Moor in these last two acts, a great man reduced to something like Roderigo's condition of being gulled. Tragic pain transmutes in the auditor into an agony of sheer embarrassment. If we forgive Othello as a murderer—as we surely do—the man we forgive is Emilia's 'Oh Gull, oh dolt . . . Oh thou dull Moore'—words wonderfully right, but not good for the great General's dignity.

Conventional accounts of the tragedy often draw the moral that Othello lacks faith—he does not trust Desdemona enough. But perhaps something like the opposite is true too: Othello trusts everybody too much, and himself too little. There is something in this splendid, dignified, and undoubtedly loving figure that is echoed and yet innocently rebuked by Desdemona's dying words: 'No body: I my selfe'. An ageing alien, Othello craves too much a Venice that will resolve for him all his uncertainties and lonelinesses, a place of 'Great-ones'. In the last act of the play, it is hard not to believe that the Moor falls to Iago's prompting because it makes easier an instinct already strong in him. In some way, Othello *wants* Desdemona to be an 'Impudent Strumpet' because his disturbed uncertainty needs good reason for looking down on her; and this is why neither of the women's simple statements of truth can reach him—and why, too, Desdemona seems to come to half-know this, and to accept it as part of the loss of innocence in marriage: 'O these men, these men'. The Moor's real need is to retain his dignity; that is what love gives him, and what love takes away. Only when he knows that his dignity is gone at last can he accept the fact of his wife's chastity, and even then with something like uninterest. He has probably now decided to die.

Othello's end, or 'sea-mark', is achieved when he has come into the estate of Desdemona, and into what Emilia calls

> not halfe that powre to do me harm,
> As I have to be hurt.

It says something about the true nature of the tragedy that these words come at last from Iago's coarse but not inhuman mate: who sees the social world of 'Mixing' finally unmix and divide into simple absolutes, the power to do harm and the power to be hurt. In this vision of equivalence the drama looks forward to the more apparently metaphysical kingdoms of *King Lear* and *Macbeth*: a range too often denied to *Othello* itself.

3
King Lear: Loving

DURING the last eighty years *King Lear* has gradually moved into a unique position in literature. It is now established as our greatest tragedy, or even more than tragedy, a *Divine Comedy* of the modern world (whether its theology is one of Gods or of no-Gods). A few dissenting voices, all very different but all beyond being called merely eccentric, have protested at this valuation. That *King Lear* does something both profound and powerful must be clear from any attentive reading or production. Moreover, *Lear* and *Macbeth* are Shakespeare's first genuinely Jacobean tragedies (where *Hamlet* and *Othello* are Elizabethan, whatever their date). And the later plays reveal their period and even their royal patronage by a certain conscious, intellectual, scope and ambition. Such qualities do much to explain how *King Lear* has in our own time come through metadrama to a condition not far short of metaphysics. This exaltation has helped to further a prolific, sometimes far-ranging criticism, which thrives both on abstraction and on extremity. Yet doubts may remain, or the sense of something gone wrong. 'Establishment' itself, the whole question of greatness, lies after all near the centre of Lear's tragedy: 'They flatter'd me like a Dogge . . . I, and no too, was no good Divinity.'[1] The play grows blurred in the smoke of incense. There may be something to be said for reducing, for once, the scale of enquiry, and asking what—in the simplest terms—the human experience is, which the tragedy's people may be said to be engaged with.

In this I am encouraged, as it happens, by the presence of that 'Dogge' in the remark of Lear's just quoted. A dog is the least metaphysical of beings, and *King Lear* is full of them—

---

[1] Shakespeare quotations are from *The Norton Facsimile*.

the play has more allusions to dogs than any other play of Shakespeare's but two: *Timon of Athens*, which in many ways seems to follow *King Lear*, and *The Two Gentlemen of Verona*. In this comedy, possibly Shakespeare's first play, the dramatist has by an act of original genius introduced among the tangled loves of his young people Launce's adored dog— a character which really has to be acted by a real dog. Critics often define the hierarchies of *King Lear*, and speak of the way it subordinates the animal world to the human. Certainly many of its characters spend their time throwing around the word 'dog' as an instance of what is most contemptible— almost as a symbol of their power of brutal contempt. But then, they do the same with the word 'slave', which doesn't prove that Shakespeare is recommending slavery. Moreover, the gentlest and kindest of the play's people edge towards a rather different position from the 'hierarchical': Edgar pities 'the poor creatures of earth' until led to promise on Lear's behalf to make the dog 'weep and wail'. And Cordelia's compassion for her father voices itself by an interesting transition: 'Mine Enemies Dogge, though he had bit me, | Should have stood that night against my fire.' Both come near to defining love as beginning in acts of kindness even to the most despised in the animal world. This is not what Lear ever—from the play's opening to its end—understands about love: 'Why should a Dog, a Horse, a Rat have life, | And thou no breath at all?' Description of the tragedy as a journey on Lear's part from Power to Love therefore needs qualifying.

The tragedy of Lear starts with his love-test. A very great deal happens after that. But the play's course is set in this first scene of the play, and it never really changes direction. The short introductory conversation by the two courtiers, Gloucester and Kent, tells us that the old King has loved his two sons-in-law ('affected' was then a word suggesting passionate preference) with a surprising fairness reflected in the equal shares of the kingdom promised to them: an equality echoed in Gloucester's insistence that he loves in much the same way both his lawful son Edgar and his bastard son Edmund. When he appears, the King stresses again this large, in fact heroic intention to reconcile private feeling with public action. He intends to divide his kingdom between his daughters. And he will make this division according to the rival statements of love made by his daughters. Who loves best gets most.

Critics and commentators take this opening as vestigial, a mere beginning borrowed from the folk-tale source-situation. None the less, most pause to note Lear's folly, dividing their energies between one of two aspects of the situation. Lear's folly may depend on the very division of his kingdom—with, as a kind of subsection, his illusion that after any such division he will retain power of his own. Or it may depend on his unwisdom in assuming that his daughters' love can be judged properly by this public competition.

Whether these procedures of Lear's would or would not have proved wise is however an irrelevant question. For the King doesn't follow them; he does something entirely different. And that he does something entirely different seems to me of the greatest importance, because Lear's choice is a vital choice. He is engaged in that process which Shakespeare's Sonnets make seem perhaps the greatest in life, the reconciling of public acts and private feeling, which however is a distinguishing between Policy and Love. The committed life goes hand in hand with the pure motive. And France does seem to be quoting from a Sonnet when, late in this scene, he says: 'Love's not love | When it is mingled with regards, that stands | Aloofe from th' intire point.' But the King swerves from 'the entire point'. He disregards that confrontation which is the soul of Shakespeare's drama and the heart and mind of his sense of love. Lear has already, that is to say, divided the kingdom behind the scenes.

The silences in critical reactions to this first scene, and to Lear's prejudgement, suggest that his action is either ignored, or assumed to be some confusion of stagecraft, or a transition that no reader or audience will notice. But Cordelia notices. Her refusal is both profound sense and radical to the tragedy. The quasi-rationalism of the King's 'fairness' to his two sons-in-law is undermined by the fact that he has three daughters, not two (three don't balance out); and this third he loves, as he can hardly be said truly to love the others. Her name, which she herself tells us twice, is Cordelia, a name close to 'cordial' which means 'belonging to the heart', 'a comforter to the heart'; and Lear has manifestly kept for her the largest section of the kingdom. She is 'last and least', the crown and climax and the smallest and the youngest and the most loved—she embodies all the unfair irrationalities of love. But if she is to embody, she must be there. Lear's plan dislocates her, unmakes her: she has no part but to sit up and beg for that remnant of inheritance theoretically left her to plead for. For—and this is the last turn of the screw in Lear's doting tyranny—if the King's plan made any sense at all she would be competing not for what was due to her, large or small, but what was left after Lear handed out their earned/unearned share to Goneril and Regan.

This is a system in which the most loved is the least loved. Lear's 'adoration' makes nothing of Cordelia, disinherits and banishes her: 'Better thou had'st | Not beene borne, then not t' have pleas'd me better.' Cordelia's 'Nothing', that is to say, is less contention than description: Lear's love as much as his hatred makes nothing of her. She insists that to love him back requires that she should exist. All Lear can do now is to find out what it means for Cordelia to exist, and—this being a tragedy—he can find out only by her dying. Already in the first scene the last scene is in sight:

> we
> Have no such Daughter, nor shall ever see
> That face of hers againe.
>
> (1. i. 264–6)

Her face is, at any rate, the last thing that he does see.

'That face of hers' is a phrase that might be noted. The ghost of the rhyme that precedes it ('we/see'), the ring of the words themselves, make it sound like something from a love-poem. It suggests that Cordelia will leave the country without leaving Lear's heart and mind—that he will destroy her before he can banish her. The phrase in fact hints at a special aspect of this opening scene: at a new inwardness in Shakespeare's tragic writing, too often ignored by externalizing accounts of the play. The tragedy opens with the formality of an old tale, re-enacted with speed and conclusiveness. It is often assumed that this irreality bespeaks an action in itself quite unimportant—only everything that ensues is important. This is true and untrue. The irreality here accompanies dream-like depths of feeling; and it cauterizes and safeguards old Lear's response to his daughters, which is not incestuous. The King's extreme age perhaps also fulfils this function, setting him beyond masculinity in a monolithic inwardness.

At I. v. 23 the enraged King suddenly drops a brief interjection, ignored by them both, into his colloquy with the joke-rattling Fool: 'I did her wrong'. There is no proof that, as editors tend to assume, this means Cordelia; indeed the withholding of identity is part of the brilliance of the moment—Lear's inner life, the 'I' and the 'her' flickers with a violent brevity and is lost again, or turned away from, immediately after. A part of Hamlet's intense originality of character is glossed if we say that he is a Prince who soliloquizes. Lear is a King, and the whole play—seemingly so wide-ranging in political and economic and sociological realities—is also a history of his life of feeling. Nothing really happens to him, except that he learns that Cordelia actually exists. Shakespeare invents the Gloucester family sub-plot to give some action to a play which would otherwise be startlingly without it, but even that sub-plot climaxes in moments of tragic love—Gloucester's love for his son that kills him, Edmund's realization that Goneril and Regan have killed each other for him.

A character in a modern short story, one by Bashevis Singer, reflects on love:

Our God-fearing mothers and fathers lived a decent life without this slavery, and believe me, they were more ready to do things for each other than the people who are involved in love affairs, and this includes myself. Much of the love of our time is sheer betrayal. It is often hatred, too.[2]

In *Lear*, Shakespeare goes back to a time whose very antiquity is the primitiveness of the feeling he is seeking out. The brutality of the age is the 'heart grown brutal' which Yeats writes about; the play's wilful, violent, enduring characters, always calling on God or gods in their own image, are men of feeling, who might similarly be said to have fed their hearts on fantasy. Much of the love of Lear's time is 'sheer betrayal. It is often hatred, too': and Shakespeare has constructed a time, part archaic, part present, which exists to embody this truth.

The orthodox reading of *King Lear* insists that the King learns—he graduates from power to love. This is true. But it is also true that the Lear of the last scene knows few moods but the imperative, kills 'the Slave that was a hanging thee', and savages those who have most tenderly followed and served him, above all Kent. He gives his last breath to an order: 'Looke there.' In short, power and love are inextricable in Lear: the peculiar grandeur of Lear's voice in the storm and on Dover Cliff depends on the fusion in it of ruined power with ruined love; and this is why Cordelia hearing it even in their scene of reconciliation (IV. vii) weeps. But this is the factor which makes us accept Lear to the end, and call him great, autocratic and even childish as he is; he finds 'that face of hers' so unforgettable he has to banish it. He is Eros Tyrannos, 'Tyrannicke Love' in person. As such, he becomes extraordinarily elusive in moral terms, not simply because he learns and changes in the course of the play (what he learns is after all limited by the lines of his character: he is the King, and dies giving orders) but because from the opening scene he releases into the tragedy all the moral anarchy of love. Mad in the storm, he self-destructively tears off his clothes, as much 'blinde Cupid' as Gloucester; on Dover Cliff he is still, for all his increase of insight, splendidly dangerous,

[2] Isaac Bashevis Singer, 'Her Son', in *A Crown of Feathers and other stories* (1974), 259–60.

irresponsible, blessing 'luxury', 'Crown'd with . . . weeds'; defeated in war, he is fugitive, fantastic, bent on turning himself and his daughter into 'Birds i' th' Cage', 'Gods spies'. But in actuality few readers or audiences can judge him thus, in detachment, because Lear exists by virtue of an imaginative power given him by Shakespeare and matched by something responsive in ourselves that has much in common with loving and being loved: he holds the very centre of our attention. He is, as he says himself, 'like a smugge Bridegroome'; he is the King.

I mentioned earlier that Shakespeare's Sonnets make use of images of royalty in speaking of love. But these images are doubtful, ironic, or harshly negating, as in the well-known close of 87:

> Thus have I had thee as a dreame doth flatter,
> In sleepe a King, but waking no such matter.

Writing out of experience mediated through the common neo-Platonism of the age, the poet of the Sonnets makes this 'waking' the one great action of Love: waking to the end of loving, or to a newer and truer sense of its reality. This process is interestingly alluded to near the end of *All's Well That Ends Well*, in a passage that seems so close to paraphrasing the end of Lear as to argue some intimate connection with the tragedy. *All's Well*'s King speaks an elegy for the play's young and supposedly dead heroine, to the culpable young hero, in terms of 'love that comes too late, | Like a remorsefull pardon slowly carried'; of 'serious things we have, | Not knowing them, untill we know their grave'; and of 'our owne love waking' that 'cries to see what's done'.[3]

The distance between Lear's banishment of Cordelia to his second parting from her is one long 'waking'. Intensely dramatic as the tragedy is, and even over-full of event, nothing but this 'waking' happens to Lear. This gives the

---

[3] *All's Well That Ends Well*, v. iii. 57–65.

action of the tragedy a specific and peculiar character, as of one long unwinding, winding-up: 'this great world | Shall so weare out to naught'. The work is on one side exceptionally contemplative. The sub-plot of Gloucester and his sons, always said to be added to increase 'universality', is more likely to be brought in to drive the plot forward. This seems plain from the strange but interesting bifurcation of the last act, finally explicit ('Great thing of us forgot . . . where's the King?'). The rest of the play charts the way in which Lear and Gloucester gradually drift together, two aspects of ruined and essentially impotent love and power. Simultaneously, Edmund and Regan and Goneril grow together in their self-destructive amorous knot of vipers.

The haunting, impotent power of Shakespeare's old men is in the fourth Act summarized by Gloucester's leap into a void that isn't there; and by the fact that Lear's high-flying imperatives are now those of a madman. As Lear and his shadow, Gloucester, trace their love through the unmaking of their world and their will, vertigo and madness push to the centre of the play. Folly is the power of reason dislocated: and it is what comes to characterize not only Lear himself but Lear's Fool and—as the Fool weakens—Edgar's disrobed and lunatic courtier. The Fool's function through all the first movement of the tragedy is to make the King 'rise', to bring out into daylight that 'darker purpose' in him which Lear announces at the beginning of the action. To criticize the King's political misjudgement is only the surface of the Fool's role; the raw pain of his injunctions derives from their undermining of the hypocritical cruelty of Court life, which represses and distorts the animal truth of the human heart: 'Cry to it Nunckle, as the Cockney did to the Eeles, when she put 'em i'th' Paste alive, she knapt 'em o'th' coxcombs with a sticke, and cryed downe wantons, downe; 'twas her Brother, that in pure kindnesse to his Horse buttered his Hay' (II. iv. 122–6). Greasy hay cannot be eaten and so is stolen by the ostler. So is much love at heart (says the Fool sardonically) 'pure kindnesse'—impure, unkind, dishonest.

An addition by the dramatist to his source-materials, the

Fool brings with him from the Shakespearian comic world that bred him an incurable ache in the sweet tooth of love—a grim wisdom as of the detached mind in the sensual body. His natural element is the submerged life of feeling in the play, the whole current of the romantic erotic glimpsed through the political stuff of the story. 'Winters not gon yet, if the wild Geese fly that way': the mesmeric power of such phrases derives from a quality of emotional, even sensual dread and yearning acted out at some level of the drama.

One aspect of Lear's tragedy is hardly assimilable to its more idealized, more 'universal' image—is indeed perhaps the element in the play which helped to alienate (for instance) Tolstoy from it; and this is a grittily commonplace sexuality, a remorseless sexual pain. The Fool's songs and sayings find for these experiences a vividly figurative animal sub-world that, nagging and penetrating, drives them into any listener, wearing down, it is clear, the old King himself. The Fool in this way embodies that 'powre to hurt' which in Shakespeare's Sonnets attends love:

> The Codpiece that will house, before the head has any;
> The Head, and he shall Lowse: so Beggers marry many.
> The man yt makes his Toe, what he his Hart shold make,
> Shall of a Corne cry woe, and turne his sleepe to wake.
> For there was never yet faire woman, but shee made mouthes in a glasse.
>
> (III. ii. 27–32)

Commentators sometimes hopefully associate the 'fair woman' in this chant of the Fool in the storm, with Goneril, but mostly give the passage up. Its point may lie in a sexual euphemism in 'Codpiece' and 'Toe' (and 'Toe', I suspect, links with an earlier phrase of Edmund's that has become a textual crux, his threat to do what is usually emended as '*top* the legitimate', but is more likely to be 'toe the legitimate', perhaps with a vague sexual implication). The Fool's 'Beggers' ironically act out the ruinousness of love, its social inutility, its fusion of self-absorption with a stupid lack of self-interest: Lear and the Fool will shortly encounter Edgar disguised as an ex-courtier and ex-lover turned into ragged Poor Tom. Thus, the Fool's vain woman ('mouthes in a glasse') is the empty narcissism which Court love carries in its heart. These grim songs articulate for Lear the gradual disintegration of love as he knows it; for it can only be he whose wildly wilful love has in some sense made 'his Toe, what he his Hart shold make', who as a result 'cries woe', and who—like the lover in the Sonnet—turns 'his sleepe to wake'. Something at any rate Lear learns from the Fool, for it is this sexual darkness he voices on Dover Cliff. There he turns on women, in the figure of his daughters—of whom the Fool has said ambiguously 'thou mad'st thy Daughters thy Mothers . . . thou gav'st them the rod, and put'st downe thine owne breeches'—with the bitter incrimination of some abandoned lover: 'Downe from the Waste they are Centaures, though Women all above'.

During the storm, the loyal Fool nags Lear almost as an angry wife will attack a husband. But he begins to weaken as his place is taken by Edgar/Poor Tom, for whom Lear develops a fascination not altogether unlike a falling in love. Many critics have spoken of the dramatist's technical adroitness in amalgamating Edgar with Lear's group, almost as if Gloucester's son were a surrogate for his godfather Lear's absent daughter. Edgar has important things in common with Cordelia: if Cordelia's fate begins with her answer to Lear of 'Nothing', Edgar when fleeing from his father discovers 'Edgar I nothing am'. This verbal echo is only one clue to Edgar's real place in the play, a place that gives meaning and depth to Shakespeare's technical expertise.

Edgar is a strange character, whose wise but drifting nonentity presents readers and audiences with a simple problem: why does he fail to make himself known to his father until the news is enough to kill Gloucester? The answer, I believe, lies in the terms on which Shakespeare has paralleled his fathers and children, and in the kind of legend of love which the tragedy acts out. The first scene's courtly opening exchange between Gloucester and Kent makes plain that Gloucester intends to keep at a distance his bastard son

Edmund ('away he shall againe') and that the old courtier uses distance after the pattern of his courtly society, to promote subjection and to create fear and respect: a subjection that, if it has bred brutality in Edmund, has generated fear in Edgar.

The tragedy never makes much sense of Edgar's maladroit claim, late in the day, that Gloucester's adultery lost the old man his eyes. Nothing in the play works in this penal way. But moral logic the work possesses; and we do at once see, beginning at I. ii, how a system of maintaining advantageous 'distances' has bred the grossest misunderstanding between the father and the sons. Recoiling soft-heartedly from Edgar's supposed ill intentions to 'his father, that so tenderly and intirely loves him', Gloucester has a few moments earlier defined this 'tender and intire love': 'O Villain, villain: his very opinion in the Letter. Abhorred Villaine, unnaturall, detested, brutish Villaine; worse then brutish . . .'. There is, of course a difference—and the cool, comic tone of this scene helps us to remember it—between smiling malice and the irritability of the good-hearted. But in tragedy the harm done by the good may matter more than mere 'villainy'. And that ironic constrast of 'brutish villain' with 'intirely loves him' is underlined by being echoed two scenes later, where Gloucester's idea of a loving father is paralleled by Lear's notion of the kindness of a daughter:

> I have another daughter,
> Who I am sure is kinde and comfortable:
> When she shall heare this of thee, with her nailes
> Shee'l flea thy Wolvish visage.
>
> (I. iv. 328–31)

In a Court, it seems, to be 'kinde' is to be comfortable, and to be comfortable is to be terrible. To some degree or other, for Lear and Gloucester alike, love functions only as a self-deceiving motive of power.

Hence the frightened distancing always between Edgar and his father, which is present even when Gloucester has his eyes. It is subtly and deeply articulated in their brief encounter on the heath (III. iv), as Gloucester glimpses the nakedness of his unknown son babbling frenziedly behind the King:

GLOU. What, hath your Grace no better company?
EDG. The Prince of Darknesse is a Gentleman. *Modo* he's call'd, and *Mahu*.
GLOU. Our flesh and blood, my Lord, is growne so vilde, that it doth hate what gets it.
EDG. Poore Tom's a cold.

(III. iv. 141–6)

Even face to face here, father and son are dislocated; verse and prose, question and answer don't meet, aren't heard. To Gloucester's helpless gentle fastidious urbanity, his naked son is the worst of company; and there is an oblique animus implied in what Edgar says of 'Gentlemen' and the 'Prince of Darkenesse'. The moment recurs later, on Dover Cliff. At IV. i Shakespeare so locates father and son on opposite sides of his stage that Edgar simply doesn't hear Gloucester's words of shame and pain about his son, Edgar himself, so that the distance of unrecognition continues between them; and at IV. v Gloucester's blindness seems to make him only the more acutely aware of class dialect in his helper ('Y'are better spoken'). Perhaps he fails to hear the familiar tones of a son beneath the 'Gentleman' because his inner ears are listening for a fantasy, the courtly dream-son he always wanted Edgar to be and which Poor Tom earlier helplessly harks back to ('Obey thy parents').

There can be no doubt that Gloucester loves his son. When in the storm he utters the words (III. iv), 'I lov'd him (Friend) | No Father his Sonne deerer', the phrase has an accent of appalling sincerity. Yet he goes on, 'true to tell thee, | The greefe hath craz'd my wits'—and there is something of the crazed or crooked beneath the pitiful 'I lov'd him'. The love here dissolves so fast into vanity; there is a vaunt, almost a threat in 'No Father his Sonne deerer'. Edgar's worst burden is hardly his brother Edmund's thrusting hatred; he also carries the weight of his father Gloucester's love.

That weight should surely be seen as crushing him. Hence the debility of the character through the play. It comes as a surprise to realize that in terms of extent Edgar's role is the second longest in the play—so much does his meaning lie in his nonentity, his negation ('Edgar I nothing am'). At no point does gentle philosophical Edgar wholly lose an effect of impotence. It is thought-provoking that Shakespeare finally (in the Folio) decided to give to Edgar the odd and weak if haunting couplet that ends the tragedy, articulating the incapacity of 'We that are yong' to live freely or long in the world of the old and powerful.

Yet Edgar is himself a child of the Court, a 'serving-man' whose first reaction to his blinded father is to be shocked by his retinue: 'Poorely led'; and in the end he assumes knightly armour and is, it seems, to 'rule in this realm'. He lives always in Gloucester's shadow and according to his pattern. Thus, throughout the play, and especially in the storm scenes, there glint—sometimes obscured textually by the bewilderment of compositors and editors—fascinating glimpses of some high world that Edgar has left, or Tom dreamed about. Early texts record a phrase of Tom's mad babble as 'Dolphin my boy, my boy cease let him trot by (Quarto) or 'Dolphin my Boy, Boy *Sesey*: let him trot by' (Folio)—and modern editors, here as at III. vi. 72, convert 'cease/*Sesey*/sese' to an exclamation 'sessa!'. But 'Caesar' is a name one would expect a Court animal to have (Edward VII's terrier, who drew tears from the crowds as he trotted behind the coffin in the funeral procession, was called Caesar); and I suspect that Edgar/Tom sees back to their grand stable a pair of coach-horses, 'Dauphin my boy, Boy Caesar! Let 'em trot by'—or even 'Bay Caesar', the 'bay trotting-horse'. Again, Edgar in his rhyme (III. iv) momentarily assumes the role of Child Roland, but the knight mysteriously turns into the Giant, advancing with a 'Fie, foh and fum'. Commentators are interested in the source of Edgar's devils, but have perhaps failed to note an important fact: that because they derive from an anti-Catholic pamphlet of Shakespeare's time, Harsnett's *Popish Impostures*, Edgar's devils may be 'impostures' too, spurious like the motives—so the pamphlet argues—of those Papists who carried out fake cures of so-called diabolical possession to gain their own political ends. And many conservative minds among the dramatist's contemporaries similarly believed that Poor Toms themselves were no more than a confidence trick to gain money from the compassionate. As a result, Edgar's devil-haunted disguise of Poor Tom in all his nakedness has a curious ersatz quality, sharply contrasted with the pain of Lear's madness and the grinding truths of the Fool. Like the false disguise of Tom, all these spectres of power—Dauphin, Caesar, the Giant Child Roland—hint perhaps that the ruined worldling cannot become innocent simply by taking off his clothes. For Edgar there is no way out except into nonentity.

To make moral categories simple and extreme, as this tragedy has them, can be a way of directing attention away from such categories. It is clear from one glance that Cordelia is an unshakeably truthful and good woman, as Goneril and Regan are not; that Edmund has the force of a brutal ego and Edgar has not; and, having also perhaps reflected that Edmund has been unloved, and Goneril and Regan plainly unpreferred by their father, we can in some sense set aside these moral absolutes. The heart of the tragedy is that passion which moves bad and good alike; for love, like death, leaves untouched few of these powerful, authority-ridden individuals. The characters' continual calling on gods tells us less about what Lear calls 'divinity' than of what these human beings themselves make of their love. In the event, what the bad make of it is trivial. Shakespeare even daringly allows his play's last scene, with its slow intrigue and sometimes stilted grotesque rhetoric, to reflect the tedium his wicked trio finally descended into. Goneril and Regan rule their affairs with a logic from first to last mechanical, even mathematical; seeking to subordinate their father, they destroy their lover and themselves. And Edmund's being 'beloved' is in the end no more than an empty gesture which lack of principle and practice deprive of all effect ('Yours', as he says to Goneril, with glamorous bravado, 'in the ranks of death'). It is the good souls, with all their clumsiness, their pain, and their terrible sense of honour, who hold our attention absolutely. Shakespeare's tragic centre in fact lies in the suffering of the good who are caught at the crossing of love with power.

The case of the—I have suggested—almost 'battered' Edgar takes on further resonance when seen to be closely matched by the fate of Kent. Many times over in criticism from Bradley on, Kent's failure to disclose his identity has been listed along with Edgar's equally frustrating reticence with his father as some Shakespearian vagueness or confusion. But Kent's life has meaning, and his disguise has a place in it. That disguise is brought to our notice at IV. iii (Q), when Kent tells a Gentleman that 'some deere cause, | Will in concealment wrap me up awhile'; and at IV. vii Kent goes further, actually rebuffing Cordelia's suggestion that he be 'better suited' with the request that 'you know me not, | Till time and I, thinke meet'. That 'meet' or right time seems never to come for Kent; who in desperation rather, feeling his life ebbing, lifts the mask and comes 'To bid my King and Master aye good night. | Is he not here?' It is only this arrival of the loving Kent which awakens the attention of the Court from the immediate politics of Edmund, Goneril, and Regan ('all three | Now marry in an instant'). But when the King does arrive, it is with Cordelia *in his armes* (the Folio stage-direction); so that Kent exclaims 'Is this the promis'd end?'

Edgar and Albany add to Kent's words murmured phrases not at all easy to comprehend; both clearly mean something serious and sympathetic. What Kent, for his part, means is surely plain, and to deprive him of it is to rob his life in the play of its extreme if stoic human pathos. For Kent has clearly always loved Lear. To him the tragedy is, among other things, what Edgar quotes from him as 'the most piteous tale of Lear and him'. Somewhere inside the life of heroic service and high principle, as Kent's face looks out from the hood of Caius, there is a romantic who, like all romantics, lives in the thought of 'the promis'd end', the unimaginably good tomorrow when love is mutual and happiness begins. In the Gospels the good are promised salvation if they 'endure to the end'; the idealist in love lives by a similar 'dear cause', and looks to an 'end' when time will at last be 'meet'. From the play's first scene ('See better *Lear*, and let me still remaine | The true blanke of thine eie'), Kent's future has involved a reconciliation reflecting, with all innocence and impersonality, the exchange of eyes between lovers. By v. iii Kent can hardly fail to understand that to be a 'true blank' may mean nothing but arrow-wounds. With Cordelia *'in his armes'* Lear has eyes for no one else's eyes; those around are 'men of stones', 'murderers, traitors all'; he drives his faithfullest friend 'away', declares him 'dead and rotten', and at last struggles to render him the courtesy due (after all) to a stranger. When for one instant the King becomes capable of something like recognition, it excludes all the immediate past, and therefore all Kent's self-sacrifice: 'Are you not *Kent*?' Kent's 'promised end' is to be still Kent: a man incapable of not loving.

Kent's existence is precluded from mere pathos by its disguises, its silences, which at once double the sense of his pain and close it from our view. Reticence is his armour. Moreover, his life has its own dark logic. Heroically good and sympathetic as he is, Kent is as much as Gloucester and Edgar a man of the Court; the reaction Edgar chances to report, that Kent 'having seene me in my worst estate, | Shund my abhord society', is interestingly close to Gloucester's. Kent's romantic idealism springs from that feeling for Authority which, as he says himself, holds him to the King; and it brings with it a high pride and sense of honour which include minor details like a preference to stay clear of mad beggars (*All's Well*'s Lafew is the tragic Kent's comic opposite, and his kindly acceptance of the dishonoured Parolles is instructive). Hence the complex tacit bitterness of Edgar's account of the grieving Kent, which closes on that aspect of Edgar's and Kent's service most agonizing to their shared courtly sense of honour:

> *Kent* sir, the banisht *Kent*, who in disguise,
> Followed his enemie king and did him service
> Improper for a slave.
>
> (QI, v. iii. 219–21)

I have already mentioned that the dramatist's Sonnets show a similar idealization of the beloved which confers royal status: 'I (my soveraine) watch the clock for you'. But such glorification never lacks the sound of irony, and its 'royalism', entailing the hard honour-ethos of the Court (though the love-object is manifestly unlike Lear in all other respects— age, type, even gender), incurs pain and self-contempt for the self-abasement involved. Love is, in the same Sonnet 57, 'Being your slave', is living 'like a sad slave'. For the good souls in *Lear*, the Court complex of power, love and honour earns them at last a comparable indignity of 'slavery'. This is conceivably a motive for Kent's enigmatically violent rage at Oswald, a man maddening not only in his smoothness and in his lack of reverence for Lear, but because he suddenly represents to Kent that slave self which loves, and which humiliates the more worldly and honourable side of the character. Among Kent's almost comic if excruciated insults are 'one Trunke-inheriting slave', and 'Strike you slave . . . you neat slave, strike'. All in all, Kent's silence or reticence through the play, his inability as much as Edgar to 'speak out' (the phrase Arnold used of Gray) create a deep image of human love in its shame or inhibition, its self-defeating self-sacrifices, its life 'Improper for a slave'.

I have lingered over the stress on slavery in *King Lear*, because among its many importances is the fact that the theme links Kent's life to Cordelia's death. Cordelia dies like a slave: she dies, that is to say, a death which, apart from all its obvious physical horror, also debases her social dignity, her honour. It may even have been the manner of Cordelia's murder which made the end of *Lear* especially intolerable to the reverently if complexly Tory Dr Johnson. Any person of high birth in Shakespeare's time, a world not yet quite lost in Johnson's, had a right to the 'noble', relatively painless death of the axe; only the 'slave' or 'low dog' was hanged, a degrading as well as agonizing end. Cordelia's slave-death is in striking contrast to what is Queenly in her life. In IV. iii (as it is in the Quarto) Shakespeare gives us a dialogue whose only function (it was cut in Folio) was to allow Kent and the Gentleman to announce the Queen of France's royalty as in itself of beauty and value, within a certain special courtly style: 'It seemed she was a queene over her passion . . . Smiles and teares | . . . as pearles from diamonds dropt | . . . then away she started, | To deale with griefe alone.' And the Cordelia who follows in person at IV. iii (in the Folio) speaks with an exquisitely troubled but undoubted majesty:

> A Centery send forth;
> Search every Acre in the high-growne field,
> And bring him to our eye.
>
> (IV. iv. 6–8)

The lines have a mysterious splendour, a character which governs the whole area of *Lear* that extends from the storm into the 'Dover Cliff' scenes. The poetic effect here might be ascribed to the invasion of the courtly by the natural, as Lear himself is, in these lines, invisible in the tall grass of the field. The peculiar authority that results from the fusion is magnificent but very complex morally, and not easy to grasp without an unbalancing towards fanciful sentimentality in one direction or a brutalist harshness in the other. Some help comes none the less from Cordelia's speech here quoted, which opens with a description of Lear mad:

> Alacke, 'tis he: why he was met even now
> As mad as the vext Sea, singing alowd,
> Crown'd with ranke Fenitar, and furrow weeds,
> With Hardokes, Hemlocke, Nettles, Cuckoo Flowres,
> Darnell, and all the idle weedes that grow
> In our sustaining Corne.
>
> (IV. iv. 1–6)

It is interesting that most editors continue to stage-direction what most readers still find the profoundest moment in Shakespearian tragedy, Lear's entry at IV. vi, as 'Enter Lear crowned with wild flowers'. The problem of Lear's authority here, which is also the problem of tragic love in the play, is that Shakespeare's character never actually enters thus. That 'Dover Cliff'—the whole location itself—may be Edgar's invention is of minor importance. What does matter is that the wild flowers are a confection put together by modern editors, beginning with the eighteenth-century Shakespearian Lewis Theobald. The original texts of Lear IV. vi just have the King 'Enter'; and his 'This [is] a good blocke', later in the scene, strongly suggests that like all well-born Elizabethans out of doors, the King is wearing a hat, i.e. a 'block' (whether or not daisy-chained with flowers), which he politely takes off to preach, and thus catches sight of.

The point is more material than it will probably seem. Our Lear, unforgettably 'Crowned with wild flowers', is hardly Cordelia's—and therefore perhaps not Shakespeare's. Her tone in speaking of him at IV. iv and later in this last movement of the play is though tenderly compassionate also unshakeable; it holds a note never impossibly distanced from those Elizabethan governors who countenanced no infringements of majesty and who could have the 'idle' poor whipped from parish to parish. She means it when she tells Lear, 'You must not kneele'; like Kent she longs to see him as true Authority, as King and Father. Therefore in her vision of him, Lear's 'wild flowers' are weeds, some of them poisonous. Her image at least helps in the difficult task of grasping Cordelia's exact relation to her father in the second half of the play, and our relation to them both—what we are to make of their love. It is at least worth pointing out that in the scene of their reunion, the poet gave Cordelia a speech of seventeen passionate lines, which cease abruptly as the King awakes. She addresses him only, and timidly, at the doctor's urging.

The change in Cordelia as the King wakes goes beyond a mere dissolving into tenderness. Bradley's distinguished essays on the play have taught generations of readers to believe in a wonderfully wordless Cordelia. But the fact is that the Princess is by nature eloquent; it is Lear who reduces the royally poised woman of IV. iii and iv to something nearer the 'difficult', though principled and sympathetic, girl of the first Act. Of IV. vii it needs at least to be noted that once Lear wakes, the flow of speech from the exhausted and broken old man still has the power to silence the young woman. Moreover the scene, infinitely delicate as it is, makes plain in the grave interchanges how Lear does it:

LEAR. If you have poyson for me, I will drinke it:
    I know you do not love me, for your Sisters
    Have (as I do remember) done me wrong.
    You have some cause, they have not.
COR. No cause, no cause.
                     (IV. vii. 66–70)

    Lear is very old, and a King, and still a little mad. But Shakespeare at his greatest hardly writes pointlessly. And these words make it difficult for a listener not to reflect, however fugitively, that Cordelia is recognizable on two minutes' acquaintance as likely to shelter dogs and unlikely to give her old father poison to drink. There are, that is to say, simple facts which in twenty years Lear hasn't noticed about his so much loved daughter. Indeed, to be a King is perhaps in his concept of it freedom from having to burden himself with mere human details like these: the fine fantasy of 'poor naked wretches' is simply more attractive to him than real known daughters, hence the power over him of the non-existent Poor Tom. Being told about not being loving may therefore be one of the reasons why Cordelia comes to sound so desperately muffled in this scene—is reduced to negatives, or her stoic and tragic stammers, 'No cause', 'I am'. Edgar's father makes him foolish, makes him 'Nothing'; Cordelia's, who without meaning it leads Kent to an empty end, reduces his best and most-loved daughter to a stubborn stammering, and at last to silence for ever.

    Pervasive through *King Lear* is a dark and dangerous humour. So dense and self-consistent is the play, this humour may work through a single word. On Dover Cliff

(IV. v) mad, Lear asks for the password, and Edgar answers 'Sweet Marjorum'. This is the aromatic cooking-herb, Origanum, which—a dictionary will tell us—took the post-medieval form of its name, 'marjoram', by corrupt assimilation to the Latin *maior* or *major*. Lear's password therefore means 'Bigger and better'—a good password for a King; and as he says 'Pass', Gloucester, who loves 'the King my master', pricks up his ears and says 'I know that voice'. There is a poetic irony here parallel to the thinking of the Sonnets. Shakespeare's Sonnets create the autobiographical intensity of being in love, of romantic *eros* at its most altruistic and most obsessive. But in the process of being honest these poems also discover the lucidity of true detachment. Thus, love in the Sonnets grows together with solitude. Not surprisingly, therefore, the poems which seem latest use a language peculiarly ambiguous and double. Sonnet 124, 'Yf my deare love were but the childe of state'—a great poem, and close to *Lear* in feeling and substance—defines a fidelity of love that 'all alone stands hugely pollitick' (and 'politic' was not an innocent word then); and it ties up all its enigmas in the still-obscure knot with which it closes, calling as witnesses of its love the fools of time, 'Which die for goodnes, who have liv'd for crime.'

Something of this mixture of extremity and doubt attends Lear himself, both in his relation to other people and in the response he calls out from us. He is crowned, if not with wild flowers or even with Sweet Marjoram, at least with ironies: 'I, every inch a King . . . like a smugge Bridegroome'. His love for his daughters is both creative and destructive, a heroic search for meaning that both makes and breaks.

The action of the play moves from Cordelia's 'Nothing' at the beginning to her silence at the end, and Lear rules this span by virtue of his power—the last he truly retains—to translate the first into the second. That point of conclusion is marked after Lear has entered '*with Cordelia in his armes*':

> Do you see this? Looke on her? Looke her lips,
> Looke there, looke there.

Bradley proposed that Lear dies of joy in the belief that his daughter is still alive; and for upwards of fifty years criticism has split into opposed metaphysical camps asserting or refuting the point, believers against unbelievers, Heaven against the Void. What none of this does is to look at Lear looking at his daughter.

This last speech of Lear's begins with a phrase that remains an enigma, 'And my poor Foole is hang'd'. Many editors find it intolerable to think that Lear's mind could wander from Cordelia at this moment. Others are sceptical that the Fool could so abruptly be known to share a death so coincidentally similar to hers. Most therefore interpret 'Foole', which they argue could at that time imply affection, as referring to the Princess. But it seems to me—as presumably, judging by their capital F, to the compositors of the Folio too—impossible that the word 'Foole' used in *King Lear* could fail to allude to *the* Fool. Moreover, Lear's introductory 'And' cannot simply be passed over. The solution seems to be to allow that the play's confusions are not Shakespeare's but his characters'; and especially, that in the play, as in life, the powerful old may generate confusion.

The point gains importance from the fact that Lear exhibits unclarity about the Fool elsewhere. Another of the play's well-known cruces turns on the King's habit of invariably addressing as 'Boy' a man whose bitter shrewdness suggests at least middle age. The total, at times almost conjugal familiarity that exists between Lear and his Fool gives the effect that they have known each other since the Fool entered the royal household perhaps as much as forty or fifty years ago. Lear has simply failed to notice that in the succeeding half-century the Fool has ceased to be a child. Deep familiarity, failure to notice—the stance so forms the King's relation to his Clown as to explain the peculiar exacerbation, the last-ditch wit of the Fool's nagging, with almost a hint at times of the self-hatred of the unlucky in love. Like Kent with his 'master', the Fool can see himself as Lear's 'dog': 'Truth's a dog must to kennell'. The Fool is, in short, one of

Lear's 'dogs', one of his 'slaves'—which is to say, one of the play's battered losers in love. I have already mentioned the fact that Lear's fascination with Edgar in the storm has the effect of displacing the Fool, who is silent for the last sixty lines of this encounter in III. iv. After 'Ile go to bed at noone' he goes out like a light, or like a memory of some long-past love ('So true a foole is love', Shakespeare ends Sonnet 57); no reference, no slightest allusion follows.

Or rather, none until this of Lear's. The Fool's disappearance from the middle of *King Lear* seems to me one of the wonderful things in the tragedy; and it is finished, or fulfilled, by this unresolved memory in Lear himself, just before he dies. The two enigmas summarize everything the play has to say about Lear and love, and as such they in a sense make it irrelevant whom Lear is thinking of: such confusion and such inwardness are Lear's great strength and weakness. No one but Lear, we feel (old, mad, and exhausted as he was and is) could simply *lose* the Fool in the storm, as some people wouldn't lose a dog; and only a Court could fail to mention the loss. That silence works to remind readers and audiences how far the blank in the sense of real identities outside himself is after all King Lear's tragedy. He often seems hardly to know one person from another ('Ha! *Gonerill* with a white beard?')—and yet sometimes, it has to be said, this is good for the people in question. There is even something heroic or at least generous in the confusion of a Princess with a Fool, and in the recognition, below awareness, that there is a level of creation on which both meet and come together: the level that holds the dog, the horse, the rat, the Fool, and Cordelia. 'Death' (as Elizabethans liked to say) 'is a great leveller.' Thus, the culpable act of confusion turns itself round into an obscurely heroic act of remembering; and that the King should suddenly and startlingly perhaps not have forgotten the Fool after all is one of the life-giving shocks the play is full of. There is a sense of the obdurate, incorrigible selfhood in old Lear that, after a tragedy's extent of suffering, loss, madness, and forgetfulness, all at once when we don't expect it produces that inward thought of the Fool. It is this same selfhood in Lear, the 'darker purpose' of his feeling life, that has also helped to destroy the Princess: the two acts interlock helplessly, as if Lear were love. But he is looking.

4
Macbeth: Succeeding

THERE are simple things we can say even about Shakespeare's tragedies. Each, for instance, has a governing feeling, which a production ought not to violate if it respects authenticity. *Hamlet* holds together in a reflective sadness; *Othello* by its extreme tension; *King Lear* harrows us. *Macbeth* is frightening.

A legend of unluckiness has haunted it in the theatre for a very long time—broken ankles and crashing scenery reported—and these stories don't seem particularly hard to believe. Actors and even theatre staff are likely to be sensitive, quickly projecting people, and such people when unnerved become accident prone—they lose their luck, their unthinking animal ease in their relation to the environment. And to be closely involved with *Macbeth* imaginatively is unnerving. A different aspect of theatre history may connect with this. There have been interesting and remarkable Macbeths in our own time (Olivier and Redgrave were both memorable) but no great actor seems to have made the part his own since Garrick. The reasons are easy to guess. 'Great roles' are those which a great actor can use to articulate himself through—those which leave room for his own energies and his own charisma. Macbeth is not an actor. His very great qualities do not include a certain helpless generosity, the communicativeness of the actor who will, as Hamlet says, 'tell all'. Macbeth is no hypocrite (he is not an actor to that degree) but he lives in a world of secrets, which his soliloquies light up in moments of still focus: everything between is in shadow, sometimes to a degree that puzzles both actors and critics. His journey towards murder is both unerring and pathless; his decline from IV. i to V. iii has gulfs in it. There are gulfs too in the basis of the character as Shakespeare imagines it. Macbeth is at once unusually 'external', a fighter who through murder lays waste a kingdom, and unusually 'internal', a self-comprehending imagination of descent; he is a self-destruction that lies along the centre of the play like a great flare-path. Trying to hold together his life and his understanding of it takes most of Macbeth's energies. There aren't many left for such powerful congenialities as make up a great actor's part.

All this is by no means the same as saying that *Macbeth* is not good theatre. A few years after its first performance, the usefully notetaking (though not otherwise impressive) Elizabethan astrologer Simon Forman went to see it, and was greatly struck, as many in the audience both simpler and cleverer than he must have been, by two of the play's scenes, the banquet scene (III. iv) where Banquo's Ghost appears, and v. i, the sleepwalking scene. And these are, indeed, two of Shakespeare's very greatest scenes, which use maximal theatrical inventiveness to show us the depths of human beings. Yet at the same time this theatricality isn't opposed to the stage problems I have been mentioning. It makes sense that both scenes are profoundly alarming, that they involve a helplessness, a being subjected, in the play's two major characters—the implacably haunted Macbeth and his wife, trapped open-eyed behind her own sleepless sleep. There is in the tragedy a certain comparable reserve or withdrawal; it isn't easily taken over, mastered, or domesticated. At a sheerly technical level, we can say that the play shows the first signs of a Shakespearian move towards Jacobean Court-forms like the Masque. When Macbeth's always-increasing ruthlessness, yet also doubt in himself and his world, lead him to hunt out the Witches for himself, he is drawn into scenes that are in themselves a brilliant dark masque, a flat vision of the future about which he can do nothing. I spoke earlier of the Jacobean quality of both *Lear* and *Macbeth*, and we feel it even more strongly in the second than in the first. The tragedies after *Macbeth* (*Antony and Cleopatra, Coriolanus, Timon*) begin to be very different, with a powerful scepticism in them, an irony about the heroic that shows their gradual movement towards Shakespeare's last phase of romance. *Macbeth* is the last play to hold a poise, to retain this commitment to the heroic which makes the experiences of the tragic characters seem of supreme importance.

Macbeth has further Jacobean qualities. Critics often lay stress on the fact that the play is 'Jacobean' in the sense of 'Jamesian'. Shakespeare was the chief writer of the company which had now moved under King James's patronage, and was known as The King's Men; and the play reflects Jamesian interests (the Witches are an obvious case). Most specifically, the Scottish James considered Banquo to be an ancestor of his own, and the play is from one angle history radically rewritten to compliment. Yet stressing the aspect of compliment, which is in any case hypothetical, doesn't do a great deal to explain why the play is so much better than other compliments to James. We can say that, if the play is shaped towards some specific purpose, then that act of shaping (whatever its superficial cause) has had effects very striking in themselves. *Macbeth* is in a sense the shapeliest of the tragedies, in a lucid, extreme, and austere way. The older Elizabethan criteria of plenitude, of interest, of generous mass and complexity, here give way to what we can recognize as a genuine Jacobean elegance. *Macbeth* is a formidably elegant play, even using the word in an almost mathematical sense: Macbeth 'cancels himself out'.

The tragedy works to a moral design that is in itself very simple. Criticism is occasionally led to write as if the same could be said of all Shakespeare's work, but this is not true; if it could, his tragedies wouldn't generate the intense academic debate that they do. About *Macbeth* there is remarkably little disagreement. The play produces very good criticism, and all good critics see something recognizably the same. This is why its central character has attracted the categorizing title, 'villain-hero'. The phrase, though crude and in fact not very helpful, hints at the strong moral form of the tragedy; and it also gestures, though more indirectly, at the fact that this overwhelmingly powerful, alarming play is also oddly withdrawn from us—its hero fits a category. This is very much like what happens to Macbeth himself. A great soldier with powerful imaginative life gradually turns into something that can at least be named, at the end, 'this dead Butcher'. Another example will illustrate this self-defining form of the drama. Macbeth's ruthless career moves through three separate and distinct crimes. The first is the killing of King Duncan to gain Macbeth the crown. The murder is committed off-stage by Macbeth himself, and the act carries a huge penumbra of implication—doubt, guilt, fear, remorse. Before committing it, Macbeth's hesitation is so extreme that many critics have found a problem in his motivation: it is hard to see how he comes at all to his decision, if there is a decision. Yet immediately the deed is done something is released in Macbeth and we hear, only after the event, that he has stabbed also, and without hesitation, the Grooms in the outer chamber. Macbeth's second murder is the killing of Banquo: a colleague, perhaps a friend; and though hired murderers do the killing, they are joined by a third who might or might not be Macbeth, and the Ghost of Banquo haunts him. The third murder is the entirely brutal and all but meaningless slaughter by hired men of Lady Macduff and her children: an action which apparently finds no response in Macbeth's conscience at all. These three murders—the King, the friend, the mother and child—form in themselves, in the pattern they make, a history; they tell us what we need to know about the man behind them. In one sense they tell us things more and more clearly, but one of the things told thus clearly is that which is less and less clear in Macbeth himself, a moral dwindling or darkening or blankening.

There is here a remarkably beautiful shaping or design which is aesthetic and moral at once—the two are fused. The story told of Macbeth's diminishing, the more frighteningly because thus tacitly, as it were inarguably, has produced a play in itself 'dwindled'. *Macbeth* is very short, hardly more than half the length of the three earlier tragedies. Many scholars and critics believe or assume that it must have been thoroughly cut: and they are encouraged in this belief, perhaps slightly paradoxically, by some evidences that it has at least been theatrically added to—the Hecate scenes are in a different tone and tempo and seem so much for a different

(Court) occasion as to suggest a different hand. But the whole question of the play's cutting must look rather different if we can accept that an effect of truncation is highly functional. The action is 'short'; but, in addition and more importantly, it feels short. Most importantly of all, it feels short to the Macbeths too. A reviewer of a Stratford *Macbeth* during the recent past belaboured the company for not letting us see the Macbeths actually *enjoying* their Kingship more. Their lack of that enjoyment worries the Macbeths, too: 'Nought's had, all's spent, | Where our desire is got without content'.[1] Part of the play's power, its frighteningness again, is the brilliance with which it dramatizes that 'desire . . . got without content' (and it's an interesting fact that Macbeth does horrifiedly call his footfalls, as they creep towards the sleeping old King, '*Tarquins* ravishing strides'). The play realizes Macbeth's enormous imaginative reveries, his longing for some greatness he can only express in terms of blood, and it equally dramatizes not only the nothingness of what it brings to him but the nothingness it brings him to.

This very distinctive shaping of the drama along the lines of a reductive human experience can be described in a number of different ways. In one sense it is political—the drama is wrought, we may say, to show objectively as God's will rather than as mere chance the slow overcoming of the evil Macbeth's reign in Scotland by the good Malcolm. The struggle has nothing of the chanciness, the relative ambiguity with which Octavius (say) would bring down Antony in a tragedy far more equably temperate, more sceptical. In *Macbeth* Shakespeare even allows himself the defiant colourlessness of the long, tedious, and difficult iv. iii, where Malcolm and Macduff slowly prove themselves to each other in terms vitally opposed to Macbeth's. It is as if by this point of the tragedy intensities of 'character' have got themselves associated so much with Macbeth as to mean tyranny: true kingly goodness can only show itself now as a colourless virtue, a disguised power—like a forest moving. And when

---

[1] Shakespeare quotations are from *The Norton Facsimile*.

Macbeth returns to the play, his heroic qualities have further darkened into a brutal and violent rage: he is only just recognizable as the man we knew in the beginning. The change is not unlike Satan's diminishment in the course of *Paradise Lost*: Shakespeare is initiating a new, even Baroque concept of what character might mean. In short, this structural clarity can hardly be described in political terms without making those terms simultaneously moral or even theological. In *Macbeth* an ethical austerity so governs as to produce an extraordinarily clear, severe moral pattern, and it probably makes sense to think of this moral rigour as a Calvinistical one, interestingly contrasted with a kind of more open Lutheranism in *King Lear*. The Scottish setting of course makes the Calvinistical passion more appropriate. Macbeth is a man whose fate would frighten us less if he were not in vital ways discernibly free; nothing and nobody makes him do any wrong action, and every wrong action he does is self-interested. Yet despite this very adequate freedom, Shakespeare has so crafted his work that Macbeth appears indissolubly locked in it; however free we call his choice to destroy himself and others, he can never get out of it. His own selfhood is in fact his prison: which is why a part of his action in the play is to advance corruptively upon that whole great autonomous and natural world that stands outside the individual. He makes it less and less possible to wake from his nightmare because he extends his dream outward more and more omnivorously. Macbeth follows his 'blacke and deepe desires' and they take him somewhere black and deep—out of the play itself: he has no heroic death, but dies off-stage, his severed head a hard but logical reduction of all he once was.

I have already spoken of a Jacobean constructive elegance as showing itself in the great formal coherence of *Macbeth*, the degree to which it makes critics think it must have been cut because there is no wastage. It is a play of extreme economy; everything is there, and there is no way out. This elegance shows itself in another way. If *Macbeth* has always proved slightly withdrawn or resistant as a stage work, it is without doubt a formidably fine *literary* text. It is a work rewarding to critics, partly because of the intellectual clarities I have been outlining, but partly because of its stylistic quality. It shows everywhere a densely patterned, finely consistent verbal texture, of the kind that richly repays analysis. The drama has become a part of that later Renaissance culture, verbally fascinated by the wit of conflicting extremes, which we associate with (say) Donne and Webster, both men who may have learned a good deal from Shakespeare in their turn (Webster certainly did). A moment ago I quoted Lady Macbeth's wretched acknowledgement that murder had gained them nothing—'Nought's had, all's spent, | Where our desire is got without content'. Macbeth agrees with her: 'Better be with the dead, | Whom we, to gayne our peace, have sent to peace, | Then on the torture of the Minde to lye | In restlesse extasie.' One of the play's most haunting and pervasive stylistic characteristics is a speech-rhythm that constantly contracts into self-checking half-rhyming half-lines: a device, surely, that realizes the foreshortening, the terrible presentness which Macbeth forces on himself, an existence without breadth and without perspective. It matches in detail that self-reductiveness of structure which I have described already.

The luminosity of *Macbeth*'s style takes many other forms. I will give another example, from the end of III. ii. Macbeth has already planned the second phase of his murderous career, the killing of Banquo, but almost penally—though with tenderest expressions—this time keeps his intentions secret from his wife. He presents the act as a fine surprise that will delight her when she knows about it: 'Be innocent of the knowledge, dearest Chuck, | Till thou applaud the deed.' And he turns aside to lure on the evening hour of the murder of Banquo:

> Light thickens,
> And the Crow makes Wing to th' Rookie Wood . . .

In this very famous line, the epithet 'Rookie' has worried some scholars, because in England (though not apparently in

Scotland) a rook is all but the same as a crow. Therefore the line has been emended. But it is a great line for precisely this reason. As the crow makes for, and is absorbed into, the dark wood, it meets itself. There is, that is to say, in these images a faint fore-echo of materials that we meet centuries after Shakespeare in the whole world of the literature of the supernatural, of ghosts and horrors, which loves to play with doubles, mirrors, and met selves. The mechanism of mirrors has a slight unnaturalness that upsets the mind. We can't help wondering if one day there might not be either too few or too many people reflected there—either no one in the glass, or someone behind us. Such fantasy-literature has mainly developed in a world very unlike that of *Macbeth*. But there is in the dark wood's 'Rookie' a shock. Macbeth's mind is here eradicating the freedom of Nature, the need of creatures to go to a wood that holds something other than themselves. His world loses air frighteningly—even light 'thickens', as if there were everywhere only wood to breathe, and nothing to see.

Only a few lines after this, at the beginning of III. iii, murderers who have inexplicably 'thickened' like the light, increasing from two to three, stand in the same beautiful dark frightening evening air waiting for Banquo and his son to come home. When Banquo has left his horses and finally comes in, he is still—in a very 'British' way—talking about the weather when he is killed:

> BAN. It will be Rayne to Night.
> 1. Let it come downe.

This is like an echo of 'Light thickens'. While Banquo is stabbed, someone overturns the lantern, which goes out. In the darkness, good fresh rain comes down as blood.

Macbeth is a play of great violence. Set in medieval Scotland, its wars are full of butchery, witches appear and disappear on its heath, murder lets loose a flood of blood, a woman walks night after night in her sleep. These things give the play its power to alarm, particularly allowing for the refined restraint with which they are mediated to us. But on their own all these aspects of the play—its intensity, its eldritch darkness—would soon cease to work on us, would remain external. If they genuinely frighten, it is for a specific reason. Everything in the drama is finely and profoundly humanized. We feel it as Macbeth and his wife feel it—and they are a pair of recording angels, or recording devils, of exceptional fineness. We can put this more impersonally. The obviously alien and dangerous is yoked throughout the drama with the domestic, the natural, the rain coming down as blood. Nightmares can always be woken up from; tragedies are about day-mares, the things that can't be woken up from. Lady Macbeth won't wake up from her nightmare because her nightmare is the play she is inside: the whole of *Macbeth* compacts itself briefly into v. i, at the sleepwalking, and we find ourselves in the form of our surrogates the Doctor and the Gentlewoman watching a play, in which the morally sleepwalking Macbeth has turned into his wife.

Lady Macbeth washes her hands, but because she is dreaming, no dirt comes off. Remorse is for daylight; but for the Macbeths, 'Light thickens' ('Her eyes are open', the Gentlewoman says, frightened; and the Doctor answers, 'I but their sense are shut'). In the same way, the play alarms most at those points at which we see in the context of the Macbeths' existence, a whole natural human world, a world of the kind of daylit safety we depend on, going subtly and thoroughly wrong. Perhaps the most famous case of this is the curious pain and anxiety we feel when the innocent old King arrives at the Macbeths' castle with Banquo, and the two of them stop to comment on its exquisite healthy air, and the martlets that have built their nests all around the battlements (I. vi). I will give another, less familiar, almost comic example.

At the end of II. i, before his tormented soliloquy, 'Is this a Dagger, which I see before me', Macbeth gives his servant a casual order. 'Goe bid thy Mistresse, when my drinke is ready, | She strike upon the Bell'. With all its magnificence and all its horrors, *Macbeth* turns out to have the kind of startling, almost charming details that none of the other tragedies, not even *Hamlet*, really give. With Macbeth's direction to his servant we are momentarily transported into the country-house world of Shakespeare's time where people have—as now, fairly universally—some form of hot late-night drink, carefully served up for them by their hostess.

We even get to know what the drink would have been; because early in the next scene, II. ii, Lady Macbeth tells us that she has given the Grooms of the King their posset. Posset sounds a vile drink to us, being hot milk curdled with wine, but it was very much appreciated at the time, and obviously tasted better if the only alternative was—as it more or less was—cold beer. The way to think of posset, perhaps, is as the Renaissance attempt at Ovaltine. It was an essentially wholesome, comforting drink, both well-bred and good for you when you weren't well. But the Grooms' Ovaltine *makes* them unwell—it's drugged so heavily ('I have drugged their possets') that Macbeth can stab two men to death without trouble. And his own posset-bell is really the signal that it is safe for him to go in and kill the King.

Such knowledge curdles the possets even faster than the possets' wine does their hot milk. If *Macbeth* alarms, this is why it alarms: not so much because of the images of evil in the play, but because of the images of good which cease to be good. The drama is full of simple and natural things which have lost their safety. Before Banquo's murder, Macbeth says savagely, 'To be thus, is nothing, but to be safely thus'—but the immense root unsafety begins where he is. Everything in the play says this to and about Macbeth, but one of the key scenes that articulates it is what we call the banquet scene, the scene in which Macbeth has ordered everyone to come and celebrate over supper his splendid new kingship, and he has included Banquo in the invitation: 'Faile not our Feast'. The banquet scene is so spellbinding, partly because it is a fully social occasion, and we see it as it were socially, from both inside and outside Macbeth simultaneously, swayed by both sympathetic imagination and a cooler social horror. This doubleness or dubiety is focused in word-play.

Macbeth says to the Murderer: 'Banquo's safe?' and the Murderer answers,

> I, my good Lord: safe in a ditch he bides,
> With twenty trenched gashes on his head;
> The least a Death to Nature.
>
> (III. iv. 25–7)

There is here, of course, one of the play's occasions of multiple ironies. But the bottom irony is the simple truth of the statement. The play does feel that to be Banquo and dead is safer than to be Macbeth. And a large part of Macbeth is coming to feel it too, has perhaps always felt it: 'Better be with the dead . . . | Then on the torture of the Minde to lye . . .' And it is particularly easy to feel this in the banquet scene because of the peculiar tranquillity of the Ghost.

In this extraordinary scene, one of the greatest scenes in Shakespeare, everything happens with a calm that would be perfect for a surreal farce. Only this is anything but a farce. What makes it bitterly farcical is the contradictory levels of nightmare taking place in it. One level, the top level, is simple, domestic, and commonplace. The murderous Macbeths are giving a little supper-party, only it is one of those supper parties that go catastrophically wrong. Lady Macbeth tries for a long while to exert her practical social poise, before Macbeth wrecks the occasion beyond retrieval. But he is right in doing so, since he really does see a Ghost— he not only sees it, the Ghost is there, sitting in his chair; and he comes twice, with an imperturbable, unbreakable, almost happy rhythm. When Macbeth defends himself to his wife he does so in lines that borrow this curious innocence from the Ghost—more than innocence, a transparent gravity:

> Can such things be,
> And overcome us like a Summers Clowd,
> Without our speciall wonder?

The remark might have come out of one of Wordsworth's better poems, and might concern spiritual instruction by an especially noble tramp. But Macbeth's tramp is a friend whom he has employed hired killers to murder, and who is now walking about in a shadowy form covered in blood.

If the banquet scene is a great and greatly horrifying scene, this is because of what is simply human and almost farcical in it—the social occasion gone wrong, the different forms of desperation in the Macbeths, the Ghost sitting quietly in the chair. I am suggesting that the power of the tragedy does not lie only in what is heightened and strange and alien in it, splendid as all these things are: its regicide, its darkness and dawn, its metaphysical hungers in Macbeth. The play also comes home in the things that are close to hand. This is a tragedy of great evil, and a reader could hardly read well who found the hero's actions quite sympathetic. There is even something slightly problematic, slightly alienating in the straightforward badness of Macbeth, the degree to which he does things which we most profoundly want him not to do, and which we even hate him for doing. Indeed he clearly hates himself: hence the distraught rage of the last act. Yet he remains dreadfully human. As late as v. iii, several scenes after the murder of Lady Macduff and her child, he can speak of his 'poore heart', his hopeless longing for what he can't have, 'Honor, Love, Obedience, Troopes of Friends'. The 'poore heart' never stops beating while he is alive; only it grows fainter and fainter, and the cold mind's observation of this from within is the tragic process.

I use the word 'cold' here. At II. iii, in the morning after the murder of the King, the Porter raises at least a kind of laughter with his bitter 'This place is too cold for Hell'. The remark reverberates, throwing light on Macbeth's grim humanity. Throughout the first two acts, his tormented wrestling with himself holds our attention absolutely. It holds it all the more for our perception of what he doesn't say as well as what he does. His magnificent reasonings never encounter the one simple fact why most human beings do not commit murder. He fails to feel for Duncan as a fellow human being. And this failure deepens as the play goes on.

Macbeth's story is of a man who desires kingship enough to kill for it, and whose action leads him on into proliferating crimes. This desire, which is destructive—a lust mystifyingly cold—requires explanation, as Macbeth can be felt to wish to understand himself. The explanation of Macbeth most often given is that he is an 'ambitious' man. This makes obvious sense. But there are problems. 'Ambition' is in itself a fairly limited and trivial impulse, however large its ends and rewards, and doesn't precisely account for the large imaginative resonances of Macbeth's story. Moreover, the Macbeth of the first two acts behaves so little like an ambitious man as to provide critics with a lasting problem of motivation. It is more as if Macbeth moves blindly, allowing what happens to create motives gradually. Certainly he ends as a far simpler, far more brutal man than he begins—he finishes as a man of nothing but motives.

The play itself offers a concept both less purposive and less limited than 'ambition': one which at least helps to explain what it is that relates Macbeth to his fellow human beings in a fashion at once so cold, so withdrawn and yet so violently engaged. That concept is of *success*. We first meet Macbeth as a successful man. At I. iii Rosse and Angus meet the victorious warriors, Macbeth and Banquo, and they communicate Macbeth's first promotion:

> The King hath happily receiv'd, *Macbeth*,
> The newes of thy successe.

Macbeth then remembers with grave excitement that the Witches who met him on the field of battle foretold such promotion and therefore gave what he calls 'earnest of successe'. Later, reporting back to his wife, named as his 'dearest Partner of Greatnesse' in the letter she is reading when we first meet her, Macbeth proclaims of the Witches: '*They met me in the day of successe: and I have learn'd by the perfect'st report, they have more in them, then mortall knowledge . . .*' 'Success' returns once more and fatefully in scene 7, as Macbeth confronts his desire for the crown, with the old King a guest in his house:

> If th'Assassination
> Could trammell up the Consequence, and catch
> With his surcease, Successe . . .

The writing is interestingly strange here, as if Macbeth were making up intense obscure knotted runic poems which at once reveal his nature and conceal it: 'surcease' means death, the great pool of blood beneath the old man, and 'Successe' means to Macbeth something that we cannot see clearly. Macbeth's tragedy is a coldness, and that coldness is brought about by his absolute response to a concept whose vagueness in human terms is vital to what Shakespeare is doing in the play. For the Elizabethans, 'Success' was a word and concept in process of change. Its original meaning entails the mere happening of one thing after another. 'Success' begins by meaning simply the way things fall out; just as 'succession' means a son's following or inheriting from his father. But 'succession' came to hold the special sense of ascent to a royal father's throne, and 'success' similarly moved upward on the same worldly path. It was natural for Elizabethans to use 'success' widely, to cover good and bad happenings alike; if they particularly wanted to distinguish, they wished each other 'good success' or regretted 'bad success'. This neutral sense is probably still contained in what Angus and Rosse say to Macbeth, meaning that the King is pleased with the news of how things have fallen out. And when Macbeth speaks of the 'Successe' that follows the 'surcease' he is thinking first of stopping some possible retributive future from happening—he is in will inhibiting the things that follow on.

The mocking half-rhyme of 'surcease, Successe' involves different meanings though. Medieval Macbeth is in a sense a very modern character. His infatuation with the word 'success' eagerly grasps towards the sense that we use it for now. All these first-act occurrences of the concept show it caught in the drift towards a peculiar specificity that it was undergoing historically. For us, 'success' has come to mean in effect that what happens, ought to happen—that whatever goes is good; and that if a man happens to 'succeed' (in our sense) then plainly he ought to have succeeded. Our concept of success insists that fortune is fate and even morality. Because Macbeth has won a battle—a battle whose ugly vertiginous chanciness the first scenes labour to make clear— he thinks he deserves to sit on a throne; he thinks that success spells succession.[2] Every time Macbeth hears the word 'success', he thinks of murder. Thinking so, he moves from a chance victory to a progressive deformation of all natural laws, the principles of true succession. This is why—to return to a detail I mentioned earlier—a bird makes wing, alarmingly, to a wood where it meets itself; and why the Macbeths, who early in the play can think readily of having children, end the tragedy childless, successionless. For the man of success there is power but no future, no existence in the natural.

A fascinating and enigmatic feature of Macbeth's career is that he never seems to agree to do the King's murder; it just gets done. Moreover, responsibility seems to slither from Macbeth to Lady Macbeth, and then critically from her back again to him, in a way that puzzles some critics and makes them feel that the play must have been cut. Lady Macbeth actually asks him 'What beast it was' that made him get her interested in murder, and so on, when there has been remarkably little time for any such communication. The answer, I think, is that both the Macbeths slide along the channels of success, which are not entirely voluntary. They fold into their world, and the next thing they know, they are killing the King. They start to live, that is to say, in 'the day of successe', and natural succession—laws of true cause and effect—begin to fall away from them.

This is why, as Bradley pointed out long ago, neither of the Macbeths regard themselves as actually pushing a knife into a gentle old man at all; they refer to the murder by a series of runes and euphemisms, as if it were a high but

---

[2] Emrys Jones draws my attention to the fact that, for a short period spanning Shakespeare's working lifetime, the word *success* could actually mean 'succession of heirs and rulers' (*OED*, *success*, sense 5).

appalling duty. But once Duncan is dead, the play starts its descent into clarity. Macbeth knows with more and more lucidity what it is he is going to do, and he does it. This process is brought about by Macbeth's use of success as an inward criterion. The word obfuscates for him as it does now, and when the mists clear it is too late to change. Some things that are successfully done aren't worth doing, and some are nothing but vicious when they are done. The Macbeths only find out the meaning of words by getting blood on their hands. Macbeth sacrifices all past and future time, all Nature's great objective pattern of true succeeding, a complex of events quite outside the desires of the self, to a present appetite: and his life turns into an infinite present, a 'To morrow, and to morrow, and to morrow | . . . To the last Syllable of Recorded time'. There is something curiously horrible about the fact that Macbeth, having murdered to get into History, in the end can't get out of it.

Why does Macbeth get himself locked up in this way in 'the day of successe'? This is a tragedy of suggestion more than of statement, but the opening of the play is rich in its suggestiveness. A peculiarity of it is the fact that it holds remarkably few characters. A named part is not a character, though some recent productions have clearly tried to make the Witches, for instance, characters, or Angus and Rosse characters. But this is simply against the grain of a work in which even Lady Macbeth (say) doesn't have a first name, and we have to refer to her throughout as Lady Macbeth. The tragedy tends to have, instead of subordinate characters, powerful and explanatory scene-setting. This is particularly strong at the opening. Macbeth is a luminary of Duncan's Court and its chief warrior. With remarkable economy I. ii—whose oddities have made some scholars believe it to be heavily edited—tells us in shorthand what we need to know about this world and its representative, the most gloriously successful man of the hour: whose glory is, however, from the first lines of the play darkened and made even more ambiguous because waiting for Macbeth on the edge of the heroic battlefield are the three Witches.

In I. ii, in a wonderfully strange and deformed speech, the 'bleeding Captaine', the 'bloody man', describes the scene that has made Macbeth what he is. If Macbeth hungers, as we shall find out, for success and succession, for somewhere to climb to, it is because climbing upward is a way of getting out of what the wounded Captain describes. The theatre of war is, he says, 'Doubtfull . . . | As two spent Swimmers, that doe cling together, | And choake their Art'. The Captain further tells the King that when Macbeth saved the day against a frighteningly successful rival, he

> (Like Valours Minion) carv'd out his passage,
> Till hee fac'd the Slave:
> Which nev'r shooke hands, nor bad farwell to him,
> Till he unseam'd him from the Nave to th' Chops,
> And fix'd his Head upon our Battlements.

What has worried scholars about this second scene is the fact of its literal confusions—Macbeth apparently defeating two geographically far-distanced enemies on one afternoon, for instance. The scene also describes Macbeth as apparently confronting as the enemy Norwegian King's right-hand-man a Cawdor whose treachery he seems to know nothing whatever about in the next scene. But none of this matters because the Captain's theme is in any case the almost unendurable confusion of war. This confusion actually shapes the Captain's speech. The encounter of King's Man and rebel dissolves into 'He . . . him . . . he . . . him', leaving a real and rather horrible uncertainty as to who exactly unseamed whom instead of politely shaking hands among the butchery. All that clearly arises from the mazed syntax is that the winner seems to be described as 'Valour's minion', a worrying title in itself. Similarly, when Macbeth and the traitor Cawdor meet in battle, someone called Bellona's Bridegroom—whom, again, we take to be Macbeth—

> lapt in proofe,
> Confronted him with selfe-comparisons,
> Point against Point, rebellious Arme 'gainst Arme.

The confusions of these lines (editors hesitate over how to punctuate them) help to sum up what they show, all the weak, confused, yet 'heroic' violence of the scene, in which identities, genders, even syntaxes are lost in vertigo. The writing is deep in that irony which rules throughout the play, undermining and reversing meanings. The confusion and irony with which warfare is here described, and the suggested uncertainty of the outcome, become the more important in that Duncan will at I. iv extend it to all politics, all diplomacy, all human relations:

> There's no Art,
> To finde the Mindes construction in the Face.

The battlefield, that is to say, extends into the Court, as the Court into the Castle. Human relations themselves are summoned up in the image with which the Captain opens:

> Doubtfull it stood,
> As two spent Swimmers, that doe cling together,
> And choake their Art.

They pull each other down in their frenzy to survive—an image we may later think of in relation to Macbeth and his wife. The 'swimmers' suggest an irreversible pull downwards in existence, a something in human energies that self-destructs as surely as gravity pulls. This is a moral vision which is categorizable under the Calvinistic—Calvin followed the dark vision of Augustinian Christianity, which argued that without God, even human virtue is nothing but splendid vices, 'splendida vitia'. Yet a theology isn't needed to countenance this vision of will-to-corruption. It's alarming, but not surprising, that the Witches probably seem less unbelievable figures now, with the signs of revival of interest in black magics, than they appeared to the eighteenth and nineteenth centuries, when it still seemed possible to believe in a progressive movement in civilization.

Matthew Arnold called up a scene not unlike the opening of *Macbeth* when he wrote in 'Dover Beach' that

> we are here upon a darkling plain
> Swept by confused alarms of trouble and flight
> Where ignorant armies clash by night.

But there is a sense in which Arnold's image is only an epic simile, brought in to intensify the appeal to emotional relationship. Shakespeare's Marguerite is his Witches. The battle is the Court's other place, the arena in which Macbeth embodies the Court's perpetual pathless climb upward. The Witches put that climb in a moral perspective, and direct the climb upward simultaneously downward. They make sure that Heraclitus's 'Way up is the way down'.

They don't effect more than this. Supernaturally speaking, they reveal only two properties in the course of the play. They can vanish, but then so could most people with the aid of Jacobean stage-machinery. And the Witches can tell the future. Or rather, they can tell one thing about the future: 'Haile to thee *Thane* of Glamis | . . . haile to thee *Thane* of Cawdor. | All haile *Macbeth*, that shalt be King hereafter.' The Witches have the same faculty as regards human success as animals—cats, for instance—have about the smell of cooking kidneys. They can tell it from an amazing distance, and they gather. There is nothing else in the Witches except a pulse beat of minute, dangerous, absolutely incessant ill will:

> in a Syve Ile thither sayle,
> And like a Rat without a tayle,
> Ile doe, Ile doe, and Ile doe . . .
>
> Ile dreyne him drie as Hay:
> Sleepe shall neyther Night nor Day
> Hang upon his Pent-house Lid:
> He shall live a man forbid . . .
>
> (I. iii. 9–22)

None of this is how Macbeth sees it, and writes about it to his wife:

They met me in the day of successe: and I have learn'd by the perfect'st report, they have more in them, then mortall knowledge. When I burnt in desire to question them further, they made themselves Ayre, into which they vanish'd. Whiles I stood rapt in the wonder of it, came Missives from the King . . .   (I. v. 1–6)

When the Witches refer to themselves in the early texts of the play, they call themselves not 'Weird'—which is the word shared by both Shakespeare's source Holinshed and modern editions—but 'Weyward', which may be only a contemporary spelling of Weird, or may not. Macbeth sees weyward old women—trivial, grubby, infinitely malevolent, but apparently fond of animals—and he makes them 'Weird', figures of power, figures of awe.

It's conceivable that we can guess why he does this. *Macbeth* is a tragedy extremely bare in human relationship: which is what gives that between Macbeth and Lady Macbeth such intensity, such strength and life. The two Macbeths between them sacrifice every other possibility of relationship that might have opened round them. But every now and again there is a kind of curtain lifted in the tragedy and we see one of these lost possibilities. While Macbeth is murdering Duncan Lady Macbeth tells us, 'Had he not resembled | My Father as he slept, I had don't'. Lady Macbeth with a father is a troublingly different thing. The Witches have a somewhat similar effect on Macbeth. Because of his murders, he needs them—he is drawn into their company, he hunts them out, it is almost an infatuation. He is even, we might say, like a weak and childish man who is always abandoning his family and allows himself to be drawn back into the spoiling care and comfort of a collection of elderly female relatives. The Witches are dingy, bad pre-Volumnias, who locate in the play the faintest hint of that destructive power-relationship of mother and son more sharply present in *Coriolanus*: 'Give me, quoth I'. The resemblance is just enough to make us see how Macbeth lacks that human rootedness as well. Certainly these unreal 'mothers' give him nothing except a form of knowledge he would in any case have been infinitely better without.

The simplest form of the evil that the Witches do to Macbeth—though the truth is that he does it to himself—is to draw him away from Lady Macbeth. Because of the play's marked thinness in terms of character, we feel the co-existence of these two acutely. Though the soul, as it were, of Macbeth's tragedy is the slow destruction of his inner self—the luminous moral imagination that understands everything that is happening to him—the heart of his tragedy is the destruction of his marriage. Lady Macbeth is all relationship to Macbeth—she is his rooting in human life.

In a recent interview about his production of *Macbeth*, Sir Peter Hall remarked—and he said it several times, with insistence—that Lady Macbeth is 'very, very sexy'. I find this modish jazzing up of the part saddening; it almost ideally misses by overshooting the real point of the darkly ironical fact that the Macbeths are probably Shakespeare's most thoroughly married couple. Not just lecherous but married: and these aren't the same things. The Macbeths have an extraordinary community and complicity. Some of the play's most troubling moments are those which reach ahead through (say) Chekhov and Ibsen and Strindberg, and many current writers, into 'the woe that is in marriage': the Macbeths become that terrible couple who appear so early in the play, 'two spent Swimmers, that doe cling together | And choake their Art'—their love is so corrupted by the struggle to survive as to pull each other down. Macbeth, to put it simply, loves Lady Macbeth;—they love each other; at the painful III. ii, where they first show a marked drift away from each other, each minds. Macbeth addresses his wife with troubled extra care, as 'Love', 'deare Wife', and 'dearest Chuck'. The tragic mutual destructiveness of the marriage is summed up by a simple fact. Married couples invariably, if it is a true marriage, grow like each other. The Macbeths slowly exchange qualities in the course of the play. From the beginning Lady Macbeth has brought to their life a directness, a practicality, an inability to see difficulties in a good cause. Only, she can't see difficulties in a bad cause, either. 'But screw your courage to the sticking place | And wee'le not fayle.' The crux over what precisely her first 'We faile?' means is interesting: she genuinely can't imagine—she can't cope with ifs; she simply throws Macbeth's phrases back at him. And this practicality moves into Macbeth in the form of brutality—which is why he starts not to need her any more. Lady Macbeth for her part inherits his imagination, but only in the form of nightmare. And she can't live with it: it stops her sleeping ever again.

I want to stress the fact of Shakespeare's depth and seriousness and even tenderness in depicting their marriage. One of the play's most touching and subtle moments is that which brings Lady Macbeth before us for the first time, and she is reading Macbeth's letter: he exists for her when he isn't there. He exists too much for her when he isn't there, she plans and thinks ahead too much for him, she too much connives, putting her image of Macbeth's future where her conscience should be: as the Doctor says, staring at her wide-open but out-of-touch gaze, 'You have knowne what you should not.' And the Gentlewoman adds, 'Heaven knowes what she has knowne.' Lady Macbeth is Macbeth's 'Dearest Partner of Greatnesse', the tender yet arrogant phrase Macbeth uses to her in his letter, as if the one thing in the world a good marriage were for, were getting a throne. And there begins from that point the insidious corruption of the good which I mentioned earlier:

> yet I doe feare thy Nature,
> It is too full o'th' Milke of humane kindnesse,
> To catch the neerest way . . .

For Lady Macbeth's immediate action, when she knows that the King is coming, is to call to spirits to 'Unsex me here'— to make herself no more a woman. She is sacrificing to Macbeth's success his succession—their hope of children. When the two of them meet at I. vii, it is the hope of children, and the destruction of children, that is a theme of what they say to each other. There is a kind of strange sense in the fact that when finally—in v. v—Lady Macbeth dies, evidently by her own hand, Macbeth feels her death as real, yet as 'signifying nothing'. He has created a present in which there is no time for death. Because he has succeeded, he cannot grieve for the one person he cared for absolutely, the person who was in a strict and technical sense 'his life'.

I have stressed Macbeth's relation with his wife, and spent little time on his more inward and imaginative life in the play, for two reasons. Macbeth's moral disintegration is very superbly articulated in the play; it is not difficult to understand; and in any case has had an enormous amount of critical attention devoted to it. My second reason is even stronger. A problem of the play seems to me how to explain why we mind it so much—why we feel great pain at the end of this savage warrior murderer. One of the most remarkable aspects of Shakespeare's genius is the degree to which in the last act he shows the brutish and external Macbeth—'Hang out our Banners on the outward walls'—and yet retains for him a pure human sympathy, so that we can still associate his fate with our own:

> I have liv'd long enough, my way of life
> Is falne into the Seare, the yellow Leafe,
> And that which should accompany Old-Age,
> As Honor, Love, Obedience, Troopes of Friends,
> I must not look to have: but in their steed,
> Curses, not lowd but deepe, Mouth-honor, breath
> Which the poore heart would faine deny, and dare not.
>
> (v. iii. 22–8)

This is the Macbeth whom Shakespeare has created as the human creature in pursuit of success, and who feels the deepening intensities of the pain of true human failure. We see here, as at moments everywhere through the tragedy, the larger human life that Macbeth and his 'poore heart' might have enjoyed, but which he has failed to understand. This experience Shakespeare sums up in one of the last lines Macbeth speaks before he is swallowed up by his last battle—a line of extraordinary power in its simplicity, and indeed it seems to me one of the very great lines in literature: 'I 'ginne to be a-weary of the Sun.' Murder and tyranny are transmuted into altogether other human simplicities.

PART TWO
Approaches to the Tragedies

5
Romeo and Juliet: The Nurse's Story

THE heroine of *Romeo and Juliet* enters the play late. Not until the third scene of the first Act is she called on-stage by her mother and her Nurse, who are also appearing here for the first time. The latter part of this scene is given to Lady Capulet's brisk and formal announcement of an offer for her daughter, with Juliet's timid and obedient response. All the earlier part of it is dominated by the Nurse, and her reminiscences of the past set the tone for the first appearance of the only three really important women in this romantic and domestic tragedy. Lady Capulet's conventional niceties make their point too, but it is the Nurse who holds the stage. Indeed, her 'moment' seems to have an importance in the play as a whole which has not been recognized. It demands to be looked at in a little detail. At Juliet's entry, mother and Nurse are discussing her age:

LADY C. She's not fourteen.
NURSE. I'll lay fourteen of my teeth—
  And yet, to my teen be it spoken, I have but four—
  She's not fourteen. How long is it now
  To Lammas-tide?
LADY C. A fortnight and odd days.
NURSE. Even or odd, of all days in the year,
  Come Lammas Eve at night shall she be fourteen.
  Susan and she—God rest all Christian souls!—
  Were of an age. Well, Susan is with God;
  She was too good for me. But, as I said,
  On Lammas Eve at night shall she be fourteen;
  That shall she, marry; I remember it well.
  'Tis since the earthquake now eleven years;
  And she was wean'd—I never shall forget it—
  Of all the days of the year, upon that day;
  For I had then laid wormwood to my dug,
  Sitting in the sun under the dove-house wall;

My lord and you were then at Mantua.
Nay, I do bear a brain. But, as I said,
When it did taste the wormwood on the nipple
Of my dug, and felt it bitter, pretty fool,
To see it tetchy, and fall out with the dug!
Shake, quoth the dove-house. 'Twas no need, I trow,
To bid me trudge.
And since that time it is eleven years;
For then she could stand high-lone; nay, by th' rood,
She could have run and waddled all about;
For even the day before, she broke her brow;
And then my husband—God be with his soul!
'A was a merry man—took up the child.
'Yea', quoth he, 'dost thou fall upon thy face?
Thou wilt fall backward when thou has more wit,
Wilt thou not, Jule?' And, by my holidam,
The pretty wretch left crying, and said 'Ay'.
To see, now, how a jest shall come about!
I warrant, an I should live a thousand years,
I never should forget it: 'Wilt thou not, Jule?' quoth he;
And, pretty fool, it stinted, and said 'Ay'.

LADY C. Enough of this; I pray thee hold thy peace.

NURSE. Yes, Madam. Yet I cannot choose but laugh
To think it should leave crying and say 'Ay'.
And yet, I warrant, it had upon its brow
A bump as big as a young cock'rel's stone—
A perilous knock; and it cried bitterly.
'Yea', quoth my husband, 'fall'st upon thy face?
Thou wilt fall backward when thou comest to age;
Wilt thou not, Jule?' It stinted, and said 'Ay'.

JULIET. And stint thou, too, I pray thee, nurse, say I.

NURSE. Peace, I have done. God mark thee to his grace!
Thou wast the prettiest babe that e'er I nurs'd;
An I might live to see thee married once,
I have my wish.

The one detail in these rich ramblings that has earned examination is the earthquake. There were real earthquakes in England in the 1580s, and one in 1580 big enough to be long memorable; and some have hoped that the Nurse's allusion might date the play. But this is perhaps to fail to grasp the very special milieu set up in these passages. The Nurse's mind has its precision, but not one such as to make her sums trustworthy. There is even a slight oddity about the figures involving the infant Juliet, since to have been only just weaned, and to be only just 'waddling' about, at rising three years, seems backward even for rustic Tudor non-gentry babies. Mathematical computations clearly increase the Nurse's dither.

'Dither' may be said to be the point of this speech. We can look in it, that is to say, for human interests and purposes even if we cannot trust its figures; indeed, the figures may be there simply to divert us from looking for the wrong thing. The Nurse's speech is a highly original piece of writing. It is perhaps Shakespeare's first greatly human verse speech, so supple in its rhythms that its original text—the Good Quarto—prints it as prose. Indeed, this looseness of rhythm, when added to the idiosyncrasies of the thought-processes as far as logic and mathematics are concerned, has increased the suspicions of some scholars about the authenticity of the whole; suspicions which can only be met by setting forth clearly a justification for it.

In part we can explain what the Nurse says here in terms of 'character' interest. In Brooke, the main source, the heroine's old Nurse holds forth to Romeus about Juliet as a small baby, and tells how she 'clapt her on the buttocks soft and kist where I did clappe', in a moment of coarse and genial humour that Shakespeare is perhaps remembering and adapting. And Brooke too has 'beldams' who

> sit at ease upon theyr tayle
> The day and eke the candlelight before theyr talke shall fayle,
> And part they say is true, and part they do devise . . .

The Nurse is a product of this comfortable and recognizable world. Shakespeare has taken Brooke's sketch of a conventional character-type and given it a dense human solidity; moreover, later in the play the Nurse will find herself in a further dimension, a moral context that defines and painfully 'places' her. In this, her opening speech, a mere something given by the story-situation is first and most massively 'rounded out', and there are also perhaps hints of that moral context to come. Her role as Nurse, her comfortable humanity, and her limitations of vision are all revealed in the references backward to Juliet's babyhood, and in the profuse mindlessness which is the medium of narration.

On the other hand, such a character need not have been quite as comical as the Nurse: and something important is contributed to *Romeo and Juliet* by the fact that she and her counter-poise Mercutio are each, in their opposed ways, exceptionally funny. She is a 'natural' and he is a 'fool', and this fact makes a good deal of difference to the way we respond to their two 'straight men', the hero and heroine of the play. Romeo and Juliet are two romantic children, but we take them—or should take them—absolutely straight; and we might fail to do so if it were not for the obliquity, or folly, that characterizes their constant companions. That is to say, from the beginning what the Nurse has is more than personality: it is function; and by function she is a 'natural'. The presence of Bottom in *A Midsummer Night's Dream*, a companion piece to this play, serves to suggest that the discourse of Shakespeare's fools and especially of his naturals will provide insight even—or most—where it appears to be failing to provide information. There is a kind of insight early achieved in the Shakespearian comic mode which can shift the comic up and away from the limits of the satirized or satirizing and into a medium which is a form of truth; or perhaps one ought to say, which is *another* form of truth. If a fat middle-aged woman congenitally disposed to muddle is made, by function, into a fool licensed to speak profound nonsense, then she may undercut the rational and move into an area of more primitive and powerful (though more elusive and dangerous) utterance. The Nurse's speech is followed by Lady Capulet's thin and superficial conventionalities, and these latter help to intensify by retrospective contrast the crude depth achieved by the Nurse.

I would argue that the major function of the Nurse's

speech is to provide a *natural* context for the motif of 'death-marked love' which governs the play. Such intimations of mortality as occur here hardly rise to tragic dignity. But it is commonly agreed that *Romeo and Juliet* makes tragedy out of the lyrical and comical. The Nurse's jokes operate well within that region of the 'painfully funny' which comes fully and deeply into being at the death of Mercutio. Indeed, one might call Mercutio's death-scene, with the astonishing death-blow given unheralded to the irresponsibly free and funny young man, a perfect match or counter-poise in a harsh vein to what is set forth here with a rough tenderness. What the Nurse says at this early point acts as a semi-choric commentary, helping to build up the background of suggestions which in the earlier part of the play act as an unconscious persuasion stronger than the explicit feud-motif in accounting for the catastrophe. It might be objected that this would demand an audience impossibly acute, able at once to laugh at the Nurse, relish her 'character', and respond to the more impersonal connotations of what she says. But it must be pointed out that for the original theatre audience this charmingly comical account of a marriageable girl's infancy was narrated on a stage hung everywhere with black. The reference to 'Juliet and her Romeo' at the end of the play certainly makes it sound a story already very familiar, almost fabulous; but even those not familiar with the tale could hardly fail to observe that a death was likely at some point to take place: that they were assisting at a tragedy. They could not be wholly unprepared to hear, at the very least, a touch of painful irony in the lines that close the Nurse's affectionate apostrophe:

> Thou was the prettiest babe that e're I nursed;
> An I might live to see thee married once,
> I have my wish.

'Married once' just about covers Juliet's case. It seems worth while to look at the Nurse's speech in rather closer focus than it has received.

The passage falls into three sections: the first concerned

with Juliet's age and birthday ('On Lammas Eve at night shall she be fourteen'), the second with the child's weaning, and the third with the child's fall. First things first: the birthday. Lammas Eve is July 31st, and so an appropriate date (as the New Penguin editor has pointed out) for a heroine named from July. But there may be a particular resonance in the festival date, which is thrice repeated, with an effect as much of ritual as of wandering memory. The Christian feast of Lammas took the place of what was possibly the most important of the four great pagan festival days, the midsummer feast. 'Lammas' itself meant originally 'loaf-mass', the sacrament at which were offered loaves made from the first ripe corn, the first fruits of the harvest. One therefore might expect Lammas Eve to carry, for an Elizabethan consciousness, mixed and fugitive but none the less suggestive associations, both with Midsummer Eve and with harvest festival. Such associations would be appropriate. For *Romeo and Juliet* is a summer tragedy as its companion-piece, *A Midsummer Night's Dream*, is a summer comedy. *Romeo and Juliet* so consistently evokes different aspects of high summer, both inner and outer weather, that Capulet's 'quench the fire, the room is grown too hot' (at I. v. 29: apparently borrowed from the wintry season in which this part of Brooke's poem takes place) is often noted for its discordance with the general 'feel' of the play. We are told that the furious energies of the fighting, fornicating, and witticizing young men are in part to be explained by the season of 'dog-days': 'now is the mad blood stirring'. The relation of hero and heroine embodies a different, more tender aspect of summer: the lyrical sense of a time that 'Holds in perfection but a little moment' (Sonnet 15). In the balcony scene,

> This bud of love, by summer's ripening breath
> May prove a beauteous flow'r when next we meet . . .

but in the tomb,

> Death . . . hath suck'd the honey of thy breath.

Then at the end of the play, these 'midsummer' associations

are replaced by an image in which the golden statues are something much more like first fruits:

> As rich shall Romeo's by his lady's lie—
> Poor sacrifices of our enmity!

A reference to Lammas, then, may carry a proleptic suggestion both of the fall that follows the midsummer equinox in the course of nature and of the sacrificial offerings of first-fruits. And there is a further point to be made, concerning Elizabethan idiom. The expression 'latter Lammas' was used to mean 'Never'—a time that will never come. The more sombre, if tender side of these hints is strengthened by the Nurse's references to Juliet's dead foster-sister, Susan.

> Susan is with God;
> She was too good for me.

In Shakespeare's time, so pitifully small a proportion of babies born survived their first six years that this reminder of a massive infant death-rate brings closer to Juliet the whole context of fatality. Not very many years will separate the deaths of the two girls. And the Nurse's 'She was too good for me' is one way of interpreting the meaning of the destruction of Romeo and Juliet themselves, and it is one that is offered as a possibility by the play as a whole.

It would be unwise to argue, from all this, that a perceptive mind ought to take the hint that Juliet is unlikely to reach or much pass the age of fourteen: or to urge that an audience ought somehow to feel *consciously* that the ludicrous argument about the precise extent of Juliet's past holds ironical premonitions of the absence of her future. But the twice-repeated 'Lammas Eve' line holds between its repetitions the dead Susan: and the conjunction of birthday with deathday lingers in the mind. The effect is not irrelevant to a tragedy in which Juliet reaches maturity with a suddenness and brevity both splendid and shocking.

To speak of maturity here is to bring up the whole question of Juliet's age, on which the passage turns in a more than merely nominal sense. The figure 'fourteen' is obtruded upon our attention so as to make it scarcely forgettable. Shakespeare is choosing an age which makes his heroine two years younger than the already very young heroine in Brooke's poem. In both stories the age of the heroine seems to have more to do with romance than with ordinary bourgeois reality.[1] Marriage at sixteen or fourteen, let alone the Nurse's 'Now, by my maidenhead at twelve year old', cannot be taken as a reflection of ordinary Elizabethan facts of life. It may be that Shakespeare was availing himself of the notion of 'hot Italy', where girls matured far earlier than in his own cooler clime, but for that the original sixteen would presumably have served. It seems important that Capulet should give the impression that Juliet is a little young for marriage—

> She hath not seen the change of fourteen years;
> Let two more summers wither in their pride,
> Ere we may think her ripe to be a bride

and that Lady Capulet should apparently contradict this later:

> By my count,
> I was your mother much upon these years
> That you are now a maid.

Considering all this, we may say that Juliet's age is important, and that the question is brought up by Lady Capulet and elaborated by the Nurse as a way of giving a good deal of information about the play's heroine, though not exactly of a chronological kind. Shakespeare is utilizing a characteristically poetic sense of time. On the one hand (he seems to insist) there is nothing abnormal about Juliet's marriage at her present age; on the contrary, given that we are moving in a romantic world, the event is a part of a great cycle—both natural and ceremonious or customary—that occurs generation after generation. On the other hand, the choice of an age slightly young even by romantic standards

---

[1] These remarks are indebted to Peter Laslett's discussion of the relatively late age of puberty and of marriage in Elizabethan bourgeois society in his valuable sociological study *The World We Have Lost* (1965).

achieves the sense of extremity, of a painful too-soonness: Juliet is a 'rathe primrose', a 'fairest flower no sooner blown than blasted'. Juliet is so young indeed that the figure of fourteen seems to suggest a coming-to-maturity that accompanies the simple physical process of puberty itself: Juliet is at a threshold. (Such hints are paralleled in Romeo's case by the adolescent fits of passion, and the rapid change of affection, which characterize him.) That Juliet is said—with some iteration—to be fourteen, is a way of establishing that she is at an *early* age for a *natural* process of maturity. Or, to put it another way, our sense of the tragedy entails both a sharp recognition of unripeness, of a pathos and gravity recognizably childish, and an acknowledgement that the grief experienced is itself 'full, fine, perfect'.

The fact that the tragic process involves a maturation brings us back to the Nurse's speech. The first of the two incidents she recalls concerns Juliet's weaning; which we may now call, in view of that movement to maturity involved with the whole tragic action, Juliet's *first* weaning. The interesting fact about the earthquake that ushers in this first movement of the narrative is not (or not only) that several such actually happened in England in the last decades of the sixteenth century, but that in this speech one happens at the same time as the weaning. This particular specimen is a poetic and not a historical event and it takes place within a context of its own. On the one hand there is the earthquake, a natural cataclysm of extraordinary magnitude, such as people remember and talk about and date things by: something quite beyond the personal—really unstoppable: it shook the dovehouse. On the other hand, there is the dovehouse, symbol—as Shakespeare's other references to doves reveal—of mildness and peace and affectionate love; and there is the Nurse, 'Sitting in the sun under the dove-house wall'; and in the middle of this sun and shelter, framed as in some piece of very early genre painting, there is the weaning of the child. The most domestic and trivial event, personal and simply human as it is, is set beside the violently alien and impersonal earthquake, the two things relating only as they co-exist in a natural span (or as recalled by the wandering mind of a natural); and because they relate, they interpenetrate. The Nurse's 'confused' thought-processes contemplate the earthquake with that curious upside-downness that is merely the reflex of those who communicate most with very small children and who speak as though they saw things as small children see them. Her 'Shake, quoth the dove-house' has not been quite helpfully enough glossed, presumably because few Shakespeare editors are sufficiently acquainted with what might be said to a very small child about an earthquake. It does not simply mean, as has been suggested, 'the dovehouse shook'; it allows the unfluttered dovecote to satirize the earthquake, as in a comical baby mock-heroic—to be aloof and detached from what is happening to it. Thus, if the dovecote gains a rational upper hand and superior tone over the earthquake, the same kind of reversal occurs in that the weaning produces an (if anything) even more formidable storm in the small child, a cataclysmic infant rage satirized by the unfluttered Nurse:

> To see it tetchy, and fall out with the dug!
> Shake, quoth the dove-house. 'Twas no need, I trow,
> To bid me trudge.

In this last phrase a fairly simple dramatic irony and pathos will be evident. Since Juliet's marriage is the subject of discussion, it is nearly time for her to 'bid the Nurse trudge' once and for all. The situation recurs in the later scene in which the young woman shows that she no longer needs support:

> Go, counsellor!
> Thou and my bosom henceforth shall be twain . . .

and helps to bring out the different pathos of the unnatural which is also latent in the situation. And the two kinds of pathos meet and fuse when Juliet is finally forced to stand free.

> I'll call them back again to comfort me.
> Nurse!—What should she do here?
> My dismal scene I needs must act alone.

But there is faintly but suggestively shadowed under this straightforward dramatic irony a different kind of irony. Throughout this whole first-act speech Shakespeare creates a poetic medium for which the Nurse's 'muddled old mind' is something of a subterfuge as Clarence's drowning vision in *Richard III* justifies itself by the conventions of dream. Because the Nurse is stupid she stands outside what she sees, endowing it with a curious objectivity. She has no moral opinion or judgement on the events that, as she pensively contemplates them, detach themselves from her and animate themselves into a natural history of human infancy. Confused and unjudged, earthquake and weaning interpenetrate in the past, sudden event with slow process: the earthquake becomes necessary, a mere process of maturing, and the weaning of a child takes on magnitude and *terribilità*, it shakes nature. The Nurse does not know the difference; and this not knowing becomes, in the course of the play, her innocence and her guilt. She has this in common, to Shakespeare's mind, with 'Mother Earth' herself, who is similarly unaware of vital differences:

> The earth that's nature's mother is her tomb;
> What is her burying grave, that is her womb.
> And from her womb children of divers kind
> We sucking on her natural bosom find . . .

The account of the weaning is less 'muddled' than so designed as to give the Nurse impressive associations such as recur much later and in the far more famous image, 'the beggar's nurse, and Caesar's'. The Nurse, lively and deathly as she is, with 'wormwood to my dug', is Juliet's natural context, the place she starts from (and Capulet's pun is relevant here: 'Earth hath swallowed all my hopes but she;. She is the hopeful lady of my earth'). Bidding the Nurse 'trudge' is the effort, one might say, of the horizontal man to be a vertical one—the human move to surpass the mere milieu of things.

The Nurse's second anecdote adds a brief, ludicrous, but none the less shrewd comment on that hunger for verticality, the perils of standing 'high-lone'. The ironic and pathetic notes of the earlier part of the speech modulate here into something brisk and broadly comic; hence the introduction of the 'merry man', the Nurse's husband, as chief actor—a replacement of surrogate mother by surrogate father, which explains the slight fore-echoes of the relationship of Yorick and the Gravedigger with Hamlet. Yet even here there is more than the merely anecdotal. The iterations, like those in the first part of the speech, are not circumscribed by the effect of the tedium of folly; there are echoes of the wisdom of folly too.

> 'Thou wilt fall backward when thou has more wit,
> Wilt thou not, Jule?' And, by my holidam,
> The pretty wretch left crying, and said 'Ay'.
>   'Wilt thou not, Jule?' quoth he;
> And, pretty fool, it stinted, and said 'Ay'.
> 'Wilt thou not, Jule?' It stinted, and said 'Ay'.

Such iterations are as close to the rhythm of ritual as they are to tedium. And they are a reminder of the presence in this play of what Yeats called 'custom and ceremony', of the ordered repetitions that frame the life of generations:

> Now, by my maidenhead at twelve year old . . .
> I was your mother much upon these years
> That you are now a maid . . .
>   I have seen the day
> That I have worn a visor and could tell
> A whispering tale in a fair lady's ear . . .
> Now old desire doth in his death-bed lie,
> And young affection gapes to be his heir . . .

This feeling for age-old process is perhaps caught up into a casual phrase of the Nurse's, a warm appreciation of the old man's unsubtle joke:

> I warrant, an I should live a thousand years,
> I never should forget it.

Involved with the husband's repetitions, one might say, is the rhythm of an existence unchanged in a thousand years. Under Juliet's particular gift, in the action that follows, for saying 'Ay' to a situation, lies any small child's easily observed habit of hopefully saying 'Yes' to anything; and under that—so the Nurse's speech suggests—lies a resilience and resurgence in nature itself.

All in all, there is considerable density of reference in the Nurse's speech. And this density is not in itself affected by the explanation we find for it: whether we choose to talk of a tissue of inexplicit conceptions within the mind of the artist himself, or whether we like to think of it as some more conscious artistry that expects a more conscious response, does not matter. The degree of deliberation that ever exists on Shakespeare's part does not seem a fruitful critical issue: it contains too many questions impossible to answer. What one can say is that the Nurse's speech presents an image of Juliet's past that happens to contain, or that contains with a purpose, a premonitory comment on her future. It alerts and reminds the audience of what is to come as do the far more formally deployed curses of Margaret in *Richard III*. But here an interesting and important complexity occurs. Margaret's curses are choric and impersonal in function: she speaks almost as Clio, the Muse of History. But the Nurse is a character in a romantic tragedy, and approaches the impersonal only in so far as a fool may. The degree of impersonal truth in her account remains a lively question. To ask whether the natural is true might have seemed in itself a not unnatural question to an Elizabethan; for Edmund, who made Nature his Goddess, was an unnatural bastard who played his brother and his father false. Both the Nurse and her vision of things are (we might say) true but not necessarily trustworthy. It is for this reason that one may call her account 'the Nurse's story'; something that offers fascinating and rich glimpses of the centre of the play from an angle that is an angle merely. She presents the play's major subjects and events—love and death—in an innocent and natural language, that of earthquakes and weaning and a fall backward. In her first story the earthquake comes out of the summer heat randomly, but not meaninglessly, for the catastrophe has scale—is a date in nature: and so with love and death. A weaning is a stage, from milk to the stronger meat of existence; so also with love and death, if we take it that it is the death of Eros in agape, and of youth in manhood which is in question. In her second reminiscence, the old man's joke reduces the complicated interwoven events of the play to a 'fall backward': and in the phrase, a childish accident, Adam's maturing sin, sexuality, and tragic death are all involved. In the connotations of the phrase, a child's innocence and an age-old blame blend with the potent romantic and erotic myth of love and death as inseparable companions, and make it startlingly harmless: romanticism grows into 'something childish but very natural'.

Through 'It stinted, and said "Ay" ' significance and appropriateness move, as through the whole of the Nurse's speech; and they are of a kind whose resonances are not easily pinned down. The action that follows certainly pins down the Nurse: what she comes down to is a randy and treacherous advocacy of bigamy. In this light we can look back and find her account of things, for all its humanity, lacking in full meaning and dignity. The Nurse's sense of 'need' (' 'Twas no need . . . To bid me trudge') does not cover a large enough human span, and the old man's consolations (' 'A was a merry man—took up the child') are clearly slightly outgrown even by an intelligent three-year-old. And yet something remains to be said. If we find some difference between the vision of *Romeo and Juliet* and that of Shakespeare's more mature tragedies, this difference might be in part put down to the effective predominance in the former of 'the Nurse's story'. Her speech establishes a natural milieu in which earthquake and weaning, a fall and a being taken up so balance that the ill effects of either are of no importance; and in so far as what she says relates to the rest of the play, it helps to suggest that the same might be true of love and death. And there seems to be a peculiar echo of her procedure in all the rhetorical doublings and repetitions of the play and especially in the paradoxes of the love and death speeches. The play's structural doublings, too, are curious, and perhaps deserve to be more often noted than they are. Romeo loves twice, once untruly and once truly; Juliet dies twice, once untruly and once truly. In any such doubling there is a point of contrast (the first love and death were illusory, the second real) but there is bound to be in implication a point of similarity also: if the first was mere game, so may the second be. Whatever the relation of the two in terms of logic, when acted out the doubled events create an imaginative equivalence.

This sense of a final equilibrium in which there is recompense for loss is in fact established as early in the play as possible, in its Prologue: which closes its doubling and paradoxical account of the feud with

> The which if you with patient ears attend,
> What here shall miss, our toil shall strive to mend.

By the end of the play it is possible to have a stubborn expectation, against all rationality, that love and death are going to 'cancel out', that Romeo and Juliet have been merely 'Sprinkled with blood to make them grow'. The image is horticultural (and is used by Bolingbroke at the end of *Richard II*). Such an image is not wholly inappropriate to a play in which Romeo lightly accuses the Friar of telling him to 'bury love' and the Friar sharply answers

> Not in a grave
> To lay one in, another out to have.

The expectation that the young lovers will 'rise again' is fairly equivocally met. Their survival owes more to art than to nature: they are no more than golden statues. Yet *Romeo and Juliet* is one of the first of Shakespeare's many plays whose peculiar quality is to make distinctions between art and nature seem false: 'the art itself is nature'. It is perhaps no accident that Mercutio and the Nurse, the play's fool and natural, turn out to be the most fertile of storytellers.

6
Hamlet: A Time to Die

'WHY does Hamlet delay?' The question has been asked for over two hundred years now. And whether or not it is the best way to interrogate the play, it seems now a natural one. For after every new reading or performance, it's difficult to avoid that prickling, sympathetic, and exasperated sensation which formulates itself as: '*Why* does Hamlet delay?' Whatever more correct form the enquiry takes, something to do with time does seem to be at the centre of *Hamlet*: which—to the extent that the play tells this kind of truth at all—makes the whole of life a great waiting game. The Prince himself, who doesn't know everything, and whose knowledge is above all that he doesn't know everything, chooses to call what is happening to him, delay; and chooses to find himself guilty of it; or finds himself guilty, whether or not he chooses; and in all these ways, may be right. The qualifications are made necessary by everything in Hamlet that makes the simple and direct 'Why does Hamlet delay?' not the best of questions, though a natural one. All questions are leading, and condition the object of enquiry in the direction of what we want to know about it. To ask this particular one is to push *Hamlet* towards presuppositions about life and literature which are in themselves doubtful, and which almost certainly didn't come into being until the period at which the question about Hamlet began to be asked: about the middle of the eighteenth century. Asking it, we suppose that men are essentially rational and motivated creatures; that in the most important issues of their lives they are free to choose; that choice is an act of will; that will and strength and success are the same thing; and that we know what we mean by *in time*. To pose and answer the question is to make of Shakespeare's play a post-Enlightenment object, whose hero will tend to be neurotic if he isn't rational, and who must be said to fail if he doesn't succeed.

We can only ask the questions we have. But sometimes defining where the questions don't fit is a way of seeing objects more clearly. And Shakespeare's play asks for, and yet sidesteps this one to a degree that is in itself informative. If the enquiry about delay breaks down, then *Hamlet* none the less answers another kind of time-question, or manifests another kind of truth about time. An analogy may help to make this point clearer. No work of art is positively like *Hamlet*; but the one most like, I think, with all its enormous differences, is Molière's *Le Misanthrope*, subtitled 'L'Atrabiliare amoureux'—'the melancholiac in love'. Half a century nearer to us than *Hamlet*, its very rationality makes its enigmas more lucid. Sad Alceste hates the world and loves its finest flower, Célimène. The question *Why*? may spring to the mind the instant the play finishes, a deference to all its direct power; but it is an evasion of that power, as well as a deference to it. The reasons either don't exist or don't matter: as for instance psychological motivation (Alceste's unhappy childhood) or biographical explanation (Molière's wife) or sociological context (the position of the playwright). Alceste hates the world and loves Célimène. *Love* and *world* are words that have very different meanings for human beings, who may find such meanings irreconcilable, without ceasing however to hunger for their reconciliation. This can be called folly or divine discontent, and may be found either tragic or comic. Alceste is in himself hardly more than a great embodiment of a linguistic difficulty, a man caught forever on one of the more usual human cruces. *Hamlet* can't be summed up like that—and perhaps *Le Misanthrope* shouldn't be either—because Shakespeare's gifts, infinitely more generous and fantastic than Molière's, were so to a degree that forfeited the later writer's economy. All the same, Shakespeare's first quite classic hero, Hamlet, is held in a crux like Alceste's, whose depth consists in its inexpressibility in terms other than its own. Explanations like, for instance, that of the Oedipus complex tend to flow off the play for this reason.

Hamlet clearly does have bad nerves. But they exacerbate, not constitute, his situation: in comparison with which a mere Oedipus complex would be child's play. A complex can, after all, be cured; that is the theory of it. What Hamlet has is both incurable and true enough to last some centuries. His situation, which is a situation in time, thus arouses that itch of assent and resentment which both recognizes the incurable and seeks to resolve it with an undermining *Why*?

Why does Hamlet delay in revenging his father? Beyond a certain point, Hamlet didn't delay in revenging his father; because he didn't revenge his father. In the end he revenged only himself. The Ghost is a presence that fades. And this fading of the Ghost is a part of the narrative of *Hamlet*, a play which offers such temporal changes and transformations as simply an aspect of the real that we know. The Ghost is first a royal presence coming to the waiting sentries; and then he is the great shadow of a loved father burdening the son with dread; and then a devil in the cellarage, friendly and bad; and finally a man in a dressing-gown whose wife cannot even see him. In the Closet Scene the Ghost stays just long enough to make us realize that we had almost forgotten him. After it, Hamlet Senior is neither present nor missed, and there is no word of him at the end of the play. The Ghost fades; and Hamlet comes into being, already a dead man. In the graveyard he says suddenly

> This is I, Hamlet the Dane
>
> (v. ii. 251–2)

taking on, *faute de mieux*, the royal title. Those who survive loss become the dead person; Hamlet ages on-stage before us, slowing down into the tiredness of 'Let be'. The Prince never does revenge his father; he does something more natural and perhaps more terrible, he becomes his father. He kills Claudius when he recognizes the similarity: 'I am dead, Horatio.'

These events may be given, at will, a psychological explanation: as when one says that a survivor will come to embody the person loved and lost. No doubt a reader so minded could retell the Ghost's story in terms of the culpability of memory: 'Why did Hamlet forget?' But there are other and similar narratives within the play which don't so immediately resolve into psychological process. *Hamlet* contains also the history of Fortinbras, whose travels to and fro through the play delineate a net that tightens around the Prince: 'How all occasions do inform against me.' The value of the golden young soldier, all-absorbed in his hunt for honour, lies perhaps mainly in the aching envy he arouses in Hamlet, along with an ironic scepticism. In comparison with that definable military glory, Hamlet's intellectual power dwindles to a single point of acknowledgement: a point that measures how much space Fortinbras has crossed since the play began, and how much time Hamlet has merely lived out. But if Hamlet can understand Fortinbras, he can't stop him—indeed the unstoppability is what he understands: Fortinbras will be there at the end, to inherit the kingdom. The fading of the Ghost, the slow and yet sudden approach of Fortinbras—diminishing Hamlet, overtaking him—are equally the play's rendering of certain great natural and impersonal laws which the individual apprehends but hardly governs. They constitute the real form of the tragedy, that quality of elastic and problematic life which makes the work seem to shift in our hands, but to shift around a centre: the curve of its own plot. The process can be given many true names, but the play leads us unusually often to think of it in terms of the laws of time, and to call *Hamlet*, if we wish, a Time Play. A human experience of living impossibility, Hamlet's crux, can be paraphrased into a very old and frequently recurring philosophical conundrum. This probably came to Shakespeare, though it hardly needed to come from anywhere, from Montaigne's fluid and incessant awareness of a fluid and incessant temporal medium. Its earliest spokesman seems to have been the Heraclitus who said 'You cannot step twice into the same river; for other waters are ever flowing on to you'. The same Heraclitus also said that 'Time is a child playing draughts . . .', and that 'In thirty years a man may become a grandfather'. The amount of time that actually passes in *Hamlet* is debatable, an issue I shall return to later. It is short enough for a child's game of draughts, the courtly tit-for-tat that is revenge; it is long enough for a man to become, if not a grandfather, at least much like his father, the King of Denmark, and dead. To neither stretch of time—the game or the growing—does the notion of *delay* seem appropriate.

It is the first of these three sayings of Heraclitus that is the best known of his utterances: and what has made it, I think, so permanently available is the absoluteness of that 'You cannot' with which the phrase begins in English. The sentence makes a natural epigram for an experience of negation radical to human life, and does so the more believably for the lack of any doctrine. An outdated physics may do very well for wisdom. Shakespeare's tragedies are founded on this acquaintance with impossibilities, and *Hamlet* more than any. At every turning in it, in phrase and image and situation, we meet an inbuilt 'You cannot', a kind of wholly intrinsic 'No Road' sign. And these negations can't in the end be resolved into inhibition, some private trouble of Hamlet's: they are, like the sayings of Heraclitus, laws of nature, statements of human physics. This is why they tend toward, and may be treated as, principles of time; and why a play which is about so many things may none the less be thought of as a Time Tragedy.

Time is a concept that we can often only hold by exercising a kind of extreme reach of intellect: and yet it is also true that the sense of time is something that immediately disappears under the impress of consciousness. When we say *Time*, time stops. Dürer's remarkable engraving of that beautiful, wild-eyed Melancholia who is Hamlet's patron saint—a melancholia that is almost the Renaissance word for consciousness—has, up in the right-hand corner, an hour-glass, and the sand is running, and yet also (because we look at it) is completely still. Shakespeare made of his revenge-plot something like the pictured hour-glass. He changed his sources to give father and son the same name, Hamlet old and young: or perhaps he took his brilliant device from the imputed early *Hamlet* of Kyd, whose static rhetoric, violent action, and obsession with memory may have become debts that Shakespeare owed gratefully. In Shakespeare's *Hamlet*, whether or not elsewhere, the revenge-plot becomes a mirror-image reversal from generation to generation: Hamlet is killed, Hamlet kills; the King is dead, long live the King. Behind *Hamlet* lie the two interlocking cycles of the Histories, from weak Harry to strong Harry, from the son to the father, from the loser to the winner and so *da capo*. The plot of *Hamlet* is a résumé of the Histories, or of history, just as the play-within-the-play is a résumé of *Hamlet*: an inset which, by the mirror-image logic that governs the entire play, opens out at the heart of its curious individual fable into a grey vista of commonplace history:

> The great man down, you mark his favourite flies;
> The poor advanced makes friends of enemies.

The plot of *Hamlet* maintains this strange static fixity, as of an object balancing its mirrored image. From the opening to the Play Scene Hamlet is Revenger, a mere function or shadow of that ghost-father whose appearances frame this half of the play. The dynamic centre of the tragedy lies at that point where Hamlet, with a gesture like that Milton was to give to the tempted Eve in *Paradise Lost* ('she plucked, she ate'), waves his arm and impales on his sword—as a pig on a spit—the hapless Polonius. From that vital and ludicrous moment to the end of the play Hamlet is Revengee, an introverted virtual image, a shadow of a shadow. And by this passive and humiliating function he achieves his end, or what seemed the end when he began. Living forwards in time, making history, here becomes what Eliot was to call in 'Gerontion', 'variety in a wilderness of mirrors'. For to any steady consciousness of the situation, the play stands still. 'Denmark is a prison'; the hour-glass does not move. Everything in the Court is a frozen shifting, an endless descent of sand: like the movement of that wonderfully distinguished speech in which Claudius sees the laws of his own and his kingdom's dissolution. In the kingdom of power and will,

> That we would do
> We should do when we would; for this *would* changes
> And hath abatements and delays as many
> As there are tongues, are hands, are accidents;
> And then this *should* is like a spendthrift's sigh
> That hurts by easing. But to the quick of the ulcer:
> Hamlet comes back; what would you undertake
> To show yourself indeed your father's son
> More than in words?
>
> (IV. viii. 118–26)

In Claudius's dreamy transition, the only paternal bequest is the will to kill. 'Hamlet comes back' in time to this.

Hamlet comes back *in time*. Similarly, Claudius lives 'when we would'. He inhabits a moment of existence conceived by Shakespeare as the time of the politicians. For, ten years after *Hamlet* was written, Claudius's peculiar melancholy, which gives this speech its heavy truth, recurs in the long silences of Alonso. Alonso is a more potent figure than Claudius: there is time and space for him, in the island after the tempest, to define absolutely the futility and the frustrations of the power-hungry will, as he sits withdrawn among the twittering time-passing courtiers. Claudius's more active capacities are embedded in Antonio. And it's an oddity of the temptation scene in *The Tempest* that Antonio seems to have his own time and space; he measures the journey to Tunis, which he left only days earlier, as light-years in length. The fact is that neither past nor future exist for him, though they have for Prospero an intensity of truth to which the play's second scene gives witness. Antonio lives only 'when we would', in the dream-like or posited moment of pure will. He is a man whom the great sea has vomited up

> to perform an act
> Whereof what's past is prologue, what's to come
> In yours and my discharge.
>
> (II. i. 243–5)

This abstract state juggles images of time, but the metaphysic is commonplace. It can be met in the weird mind-changes of any genuinely political personality, unconvinceable that the loyalties or opinions of yesterday have any meaning: too weak, that is, to maintain a truth undermined by the shifts of existence:

> It is not very strange, for my uncle is King of Denmark, and those that would make mouths at him while my father lived give twenty, forty, fifty, a hundred ducats apiece for his picture in little. 'Sblood, there is something in this more than natural, if philosophy could find it out. (II. ii. 358–64)

Hamlet's last remark here hesitates between naïvety and a sophisticated irony, as so many of his sayings do. It is both strange, and not strange at all, that courtiers condition their minds to their profit: the matter is both 'more than natural' and very natural indeed. Beyond a certain point, philosophy won't help Hamlet. Similarly, the time-worlds of *The Tempest* don't need metaphysics to explain them. In substance they reflect simple morality; in method they operate merely by Prospero's magic; and both moral and magical systems demand no more than a generous attentiveness from an audience. Dr Johnson once disposed of the unities by saying that we always know in any case that we're in the theatre: a right instinct though a wrong argument. Anyone who understands any work of art knows as much as he needs about the strangenesses of time and space. The symbolisms of art demand a natural acquaintance with conventions, an easiness in taking up and laying down relationship with the actual, that makes of any audience a Prospero in itself. Thus the Chorus of *Henry V*:

> 'Tis your thoughts that now must deck our kings,
> Carry them here and there, jumping o'er times,
> Turning th'accomplishment of many years
> Into an hour-glass.
>
> (Prologue, 28–31)

It's easier than it once was—perhaps almost too easy—to turn that hour-glass, sixty years after the advent of Modernism, which showed that the convention of space in a painting comes to little more than the relation of one brush-stroke to another, and that time is after all only a dimension. The novel started as the story of a lifetime, and George Eliot said that *Romola* took her from youth into age; but a novel doesn't take much longer to read than a twenty-line poem, also perhaps the transcript of a lifetime. Drama, particularly when staged, has a peculiar vraisemblance that can make it seem sensible to talk, as several generations once did, of time-schemes, of calendars of days and weeks and months: so that one ought for instance to be able to prove that Hamlet's time is or is not 'enough', that he does or does not *delay*. But even drama, in so far as it is art, is a mental act, and therefore almost timeless, its running-time a metaphor for as long as we like. In Peele's *Old Wives' Tale* an old woman tells a story and its characters walk on-stage and are the play. To ask 'When?' would be as misplaced as Laertes's characteristic 'Oh, where?' when he hears his sister's drowned. Peele's people could only say, like Shakespeare's Sir Toby, 'We did keep time, sir, in our catches'. More formally, one could say that Peele's work balances on a point somewhere between the aesthetic sophistication rediscovered in Modernism, and the simple crude fact that in England in 1590 no clock told the right time. For, if it is Time that is in question, we can't estimate, I think, the subtle but radical changes that four hundred years have brought to the human mind. Queen Elizabeth had a wrist-watch: it was a suitable present for millionaires and monarchs. Clocks became familiar in England only after about 1650; not until the eighteenth century did they become accurate; and absolute accuracy was the product of Greenwich Mean Time. Elizabethan villages lived by the church-bell at their centre. When we think about *delay*, we do so from a standpoint divided from the Tudor by the age of the satanic mills and factories, when clocks became what human beings clocked in by; and the word *delay* carries locked within it the implicit noise of the machine.

Shakespeare's plays time themselves, as country people did for many centuries, by the sun, and he starts up indoor clocks only as and when he pleases. *Henry IV*, Part I opens with Hal rebuking Falstaff for defining time in a way that unmakes clocks; and the play is Falstaff's in this sense, revolving around its great clock-stopping comic occasions, Eastcheap nights before the morning after. Part II by contrast belongs to the King and the old Lord Chief Justice, the action dragging itself out in the long dreary shadow of *too late*. Another example: in the middle of *Twelfth Night* Olivia is with finality turned down by Viola. There is—one presumes—a sudden off-stage bell-note and Olivia says 'The clock upbraids me with the waste of time'. The death of a fantasy, the pang of a real despised love, are a simple and hard slap in the face strong enough to start up a clock in Olivia's mind. The ghost of Hamlet's father enters at just such another stroke, 'The bell then beating *One*!' And Hamlet himself says '*One*!' like an echo of the clock, as he gives his first blow to Laertes, to whom he has said 'I loved you ever' and who in response kills him.

The word *One*, used in this way, in fact occurs with weight a third time in *Hamlet*. At the beginning of the last scene, whose packed action so crystallizes into the ritual of the duel as to seem to stand still, to be timeless one might say, Hamlet absently reassures the anxious Horatio:

> It will be short: the interim is mine,
> And a man's life's no more than to say *One*.
>   (v. ii. 73–4)

The phrase is definitive though enigmatic. Its resonance has something to do with those two other bell-notes to which it looks forward and back, the clock that strikes at the Ghost's entry and the call of the first blow struck in the duel. The three '*One*'s hold together like a linked irregular chain, or like the echoes of a stone going down a deep place. The '*One*'s can seem echoic in this way because the first of the sequence, the bell-note that brings in the Ghost, encourages the image, and is, moreover, itself an echo. The entry of the Ghost is an extraordinary dramatic event that helps to define something

of what one might mean by calling *Hamlet* a Time Tragedy.
And it begins with a man remembering:

> Last night of all,
> When yond same star, that's westward from the pole
> Had made his course t'illume that part of heaven
> Where now it burns, Marcellus and myself
> The bell then beating *One*—
>
> *Enter Ghost*
> (I. i. 35–9)

This first appearance of the Ghost is perhaps the most quietly startling moment in all Elizabethan drama. The Ghost brings to their feet the intent, seated group of listening soldiers, shattering their circle as it interposes itself within Barnardo's very sentences. It brings its own ending to his unfinished story, a subject relegating his own unspoken object to nothingness: it leaves 'Marcellus and myself' suspended forever in the air, just as it will the next night dislocate Hamlet's life, leaving thirty years unfinished for always. The Ghost's effect is even more radical than this, more transforming and more philosophical: it alters the temporal dimension of the moment. The story that was begun has created on stage that rapt enclosure of the historic where all narratives take place, a security reflected in the lamp-lit circle of listeners. When the Ghost comes on, when the Ghost— unnoticed through the familiar conjurors' distraction of spell-binding anecdote—suddenly *is* on, that security is broken; narrative is drama and the past is present. Theatre and metaphysics come together: the *shadow* behind the soldiers is a true *ghost* because a real *actor,* just as later in the play the Players crowd on-stage as the 'brief chronicles of the time', to deny that the past was ever anything but living. In this first scene of the play, time is a great continuum like an open stage. As Barnardo remembers, and the bell beats, and the Ghost comes and stands in the darkness, the present moment dissolves into a receding sequence of shadows, of haunted imagined nights all reaching back for their meaning to a time when 'the king that's dead' really lived. What is here and now grows, like Plato's shadow of a shadow, dependent for its reality on something else, something always behind it to which it traces back in a long tangle of source-relationships. The Player King will say:

> Purpose is but the slave to memory.
>
> (III. ii. 183)

But the Ghost says it first, in silence. For the whole of the rest of the play it will be foolish to think that the past is past, and dishonest to withhold sympathy from a man who finds it impossible to live wholly in the present.

The first appearance of the Ghost is a brilliant stage moment, like a flash of lightning. It is, that is to say, only a moment, but it extends in the mind into concepts and reactions much wider-reaching, their outer limits being the boundaries of the play, which are perhaps the boundaries of human society itself. The moment is a definition, as of law. Barnardo has 'called back yesterday'; from now on, every hour struck in *Hamlet* is a passing-bell. History begins when the bell strikes *One*, and the prince's vocation is to give meaning to the clock, saying *One* until the Ghost is laid at last. For the play's hero is introduced to us as *Young Hamlet*, the child of the Ghost, and the first important thing that we learn about him is that he is, in his black garments, true to the darkness, and incapable of forgetting:

> Heaven and earth,
> Must I remember?
>
> (I. ii. 142–3)

The Court of Denmark lives by daylight and survives by forgetting, its time-servers drifting on the present moment as Ophelia will on the brook that finally drags her down. In such a world, Hamlet's only freedom is to follow the Ghost; to be caught into that huge individual act of confronting Time which the play summarizes as memory, or delay: a death in life, but the life, also, of all human civility, and the source from which the Prince derives his royalty. While he thus follows the Ghost, life becomes for him 'a time to die'.

This is the impasse or crux on which Hamlet rests, and it makes of him a figure not unlike that image of Melancholy which I mentioned earlier, one created at a time when the mechanical clock and the sense of History were together beginning to master Europe. Benighted and moonlit, Dürer's powerful, winged, but seated Melancholy lifts a heavy frowning head: and behind her head, next to the hour-glass, there hangs a bell, its rope drawn sideways and out of the picture, about to ring and never ringing. In just such a condition of pause, Hamlet is held: always potential and always too late.

7
Textual Readings and Reading the Text of *Hamlet*

FOR the last few years Shakespeare studies have been dominated by questions of authorship. A new poem and even a new play have been claimed. Whether or not found convincing, such attributions have a larger interest: they can raise the whole issue of authorial identity. In one of the recent disputes a bibliographer met objections that the poem proposed was not good enough for Shakespeare with the simple response that 'much of Shakespeare was bad anyway'. Given that literary scholars devote their professional lives to their subjects, it is useful to have the grounds on which they do so argued out.

But more conventional textual scholarship, not directly concerned with the authorship question, can in fact lead to these major principles of value. I should like to offer tentative resolutions of some of the many vexed cruces in the text of one play, *Hamlet*; but I want to preface these suggestions with a fuller discussion of two examples, in the hope of giving some sense of these larger issues—the sense of what significant critical presuppositions may underlie textual discussion that is scholarly in the narrower meaning of the word.

A good place to begin would seem to be the words referred to by Harold Jenkins in his New Arden *Hamlet*, one of the play's two major recent editions, as 'probably the most famous crux in Shakespeare': 'the dram of eale'. The phrase comes at the end and climax of the Prince's long meditative speech as he and Horatio wait on the castle platform for the Ghost to appear, while they overhear the noise of the Court's drunken revelry. Reflecting on the harm done abroad to the nation's image by its famed debauchery, Hamlet is led by the theme of reputation to a more inward yet logical exploration of the whole symbiotic relationship between a man and his society, between the 'particuler' and the 'generall':

> So oft it chaunces in particuler men . . .
> these men
> Carrying I say the stamp of one defect
> Being Natures livery, or Fortunes starre,
> His vertues els be they as pure as grace,
> As infinite as man may undergoe,
> Shall in the generall censure take corruption
> From that particuler fault: the dram of eale
> Doth all the noble substance of a doubt
> To his own scandle.

I quote here, and throughout, the Second Quarto (1. iv. 23, 30–8; with modernization, here as elsewhere, of *u*, *v*, and *s*), the first of the only two original and authentic texts of *Hamlet*, published in 1604–5 possibly as corrective to the corrupt and pirated First Quarto of 1602. Any modern text of the play will need to compare with Q2 the First Folio, appreciably different in many readings and sometimes plainly, sometimes arguably, more correct. But it does omit three substantial passages, of which one is the speech at present under consideration. This omission is in itself a factor of editorial theory; and it is perhaps not surprising to find the case for authorial revision now strongly argued, given that such Shakespearian revision is, along with authorial identity, probably now the most prominent issue in Shakespeare studies. Indeed it could be said that the two are one: the Shakespeare we 'know' is the critic, the reviser. In the other of the two recent editions I shall be referring to, the new New Cambridge, Philip Edwards quotes with seeming approval J. M. Nosworthy's 'The simplest explanation of this crux ['the dram of eale'] is that the sentence is unfinished, the implication being that Shakespeare lapsed into incoherence and gave up the struggle': a view assimilable into Edwards's own sense that the Folio text reflects the dramatist's personal censorship of those parts of the play with which he was, like Edwards and Nosworthy, essentially dissatisfied.

The argument interestingly reveals the importance to textual scholarship of critical judgements and preconceptions. I don't myself find the passage remotely 'incoherent'; I find it merely flawed by the error that plainly underlies 'of a doubt'. Indeed, if Shakespeare did make the decision to cut—and the desire for a peculiar reticence in his tragedy may well have prompted the excision of what was so discursive, so explicit—then the motive was, paradoxically, the very excellence of the writing here. Hamlet too brilliantly discloses that mutuality of a man and his world which the whole action of the play works to create. In that reciprocal life the middle term is the social climate of opinion (which the Elizabethan mind ennobled into 'fame' or 'reputation'): the Court is a sphere in which a man's grasp of absolutes ('grace', the 'infinite') easily deliquesces into mere behaviour, gossip, 'generall censure'. Of all who have discussed this passage, only Jenkins seems to perceive the significance, in what Hamlet is saying, of this large theme of reputation, of observation—'the generall censure'. Necessarily the 'dram of eale' lines are talking about something *seen to happen*, something observed in process.

'Dram' is a word measuring liquid weight (as well as implying the smallest known quantity); therefore 'eale' is likely to be a liquid. Linguistically, *ea* sounds may move towards either *e* or *a*: 'ale' makes some sense in the context, but 'ele' a good deal more. The word is obsolete for 'oil'. All modern editors emend to 'evil', with the exception of Edwards who retains 'eale' while finding it meaningless, and there seems little doubt that Shakespeare intended the pun of 'evil (e'il)' in 'ele'. But 'oil' was, it seems to me, his primary meaning, for a simple reason known to anyone who has cleaned clothes or other household goods. Nothing stains so badly as black oil or is harder to get out.

The dram of oil does something to the noble substance 'to his own scandle': it ruins, that is, both its own credit and that of the substance in the eyes of a viewer (there being, surely, implication in that 'his' (= its), which ties into one both corrupter and victim). To trace the process is to see that a verb must have been misread by the printers so as to emerge in the phrase 'of a doubt'; and we can already suppose certain things about this word. I have suggested elsewhere the value in textual criticism of respecting Shakespeare's compositors enough to suppose that printers' errors often argue an authorial text of some difficulty or obscurity: as, here, underlying 'eale' is the obsolete 'ele'; 'of a doubt' may therefore mask a verb of some rarity or obscurity.[1] Moreover, the lines give their own obvious information about the word. They describe a process not so much housewifely as scientific, experimental, even alchemical—the word 'noble' brings alchemy with it: the young student Hamlet is describing an experiment watched by learned observers. (There may be, I suspect, some half-conscious memory of this passage in the young T. S. Eliot's scientific tropes for criticism; and even too in the young Auden's sinister evocation in 'The Witnesses' of the sky 'darkening like a stain'.) What happens under the 'scandalous' effect of the oil can only be the spreading of a darkening, of a black stain, as in water corrupted by the infusion of a black oil. The word underlying 'of a doubt' must be a verb suggesting that phrase in its form but meaning something more like 'darken', 'shadow', probably with a metaphorical relation to human character.

This impressively climactic and as it were 'doomed' word was, I suggest, 'overcloud'. It is a word rare enough to explain the Quarto printers' uncertainty, but it occurs (the few *OED* citations reveal) in two writers likely to have influenced the play in other ways: Kyd (*The Spanish Tragedy*) and Nashe (*Christ's Teares*). But I believe that Shakespeare remembered the word from a writer perhaps more important to the play than either of these: from Montaigne, thought by many scholars to have been read by Shakespeare in manuscript while he wrote his tragedy, despite the fact that

---

[1] See Chap. 10 below.

Florio's translation of the *Essays* was not published until 1603. On page 330 of this first edition of the *Essays*, we may find, in the 'Apology for Raymond Sebonde', a passage whose entangled meditation brings us close to Hamlet's thinking at several points of the play—though Florio as a stylist only makes clear by contrast how long we ought to pause before supposing 'incoherence' in any of the Prince's superb lucidities:

But thinkes it not, that we have the foresight to marke, that the voyce, which the spirit uttereth, when he is gone from man, so clear-sighted, so great, and so perfect, & whilst he is in man, so earthly, so ignorant, and so overclouded, is a voyce proceeding from the spirit, which is in earthly, ignorant, and overclouded man; and therefore a trustles and not-to-be-beleeved voyce?

There are purely bibliographical reasons for believing that this powerful, witty word 'overcloud'—the more memorable for being re-echoingly repeated here—was so scribbled by Shakespeare as to be misread as 'of a doubt'. All editors agree that the Folio printers were right to correct the Quarto's curious 'friendly Fankners' (II. ii. 450) to 'French Faulconers'. Clearly, the arch and pronounced descender of one form of secretary *h* in 'French' could be misread as a scrawled *y* (a fa⁻ᵗ I shall return to in a later crux); and the preceding *c* plus the first stroke of the *h*—in effect, *cl*—could be misread as *d*. On these grounds, and further given the common *t*/*d* confusion in secretary hand, *cloud* could be misread *dout*, a frequent Elizabethan spelling. Once 'dout' suggests itself the ear easily converts 'over' to 'of a'.

But there are stronger reasons for believing that Shakespeare wrote

>   the dram of ele
> Doth all the noble substance overcloud
> To his own scandal

—and they are literary rather than literal. The image of overclouding has peculiar decorum in *Hamlet*, where Claudius's first (interrupted) sentence to his difficult dark-robed stepson finishes 'How is it that the clowdes still hang on you'. Laertes too is later in the Quarto described (IV. v. 89) as keeping 'himself in clowdes'. The play's night-scenes, its hauntings, and all its panoplies of death add to what is essentially bewildered and doubtful in its substance to give especial point to that conclusive and as it were self-sentencing verb, *overcloud*—spoken surely just as the Ghost moves on to the stage (perhaps not unlike the arrival of black Marcade in the early comedy: 'Worthies, away, the scene begins to cloud . . .'). The word, like the whole speech, is powerful and original enough to rescue Shakespeare from the notion of one who 'lapsed into incoherence and gave up the struggle'. Shakespeare's coherence is triumphant and his sentence closes like a lock.

My argument has been that even the more conventional textual scholarship challenges the reader with large and implicit suppositions as to the nature or identity of the poet concerned; as, in the new New Cambridge *Hamlet*, Shakespeare is a writer first revealing his uncertainties then almost academically correcting them. The second case I want to turn to, as throwing up these interesting preconceptions, is perhaps rather less well known than the first. Early in the play's very remarkable first scene, as the Ghost appears to the amazed guards on the castle's gun-platform, one of them asks Horatio 'Is it not like the King?' and receives the answer:

> As thou art to thy selfe.
> Such was the very Armor he had on,
> When he the ambitious *Norway* combated,
> So frownd he once, when in an angry parle
> He smot the sleaded pollax on the ice.

So lines I. i. 60–4 in the Quarto; the Folio reads the last line 'He smot the sledded Pollax on the Ice'. Although their 'sleaded/sledded' differs, both Quarto and Folio are firmly agreed on 'pollax' (pole-axe). Later referred to as 'Martiall' and 'armed', the Ghost clearly carries a battle-axe. Yet almost all modern editors, including both Jenkins and Edwards, insist that by this spelling both Quarto and Folio must intend 'Polacks' or Poles, and that these warriors necessarily sit in sleds.

The arguments for these now unavoidable, almost mythical 'sledded Polacks' are multiple and various, but all are based, or so it seems to me, on a preconception voiced by Jenkins in his splendidly rich and absorbing New Arden edition. The sledded Polacks have (he says) that 'power to stir the imagination, which a pole-axe so signally lacks'. This recognition of the part played by imagination in editing, as in all reading processes, can only be welcome. And certainly, too, those sledded Polacks on the ice have irrepressible glamour: the tragic fate of Eastern Europe, the frost-fairs on Shakespeare's Thames, his Jacobean King's daughter's exile ('the Winter Queen') from Bohemia, the long dying of late-Renaissance heroic culture, all confusedly work together with other such associations to make the mystery of the line.

But did Shakespeare write it? 'Imagination' isn't a matter of just the romantic and nostalgic, but of a profound apperception of human life and character: of the likelihood of things. Hamlet's father is here being evoked as an old warrior who embodies a certain kind of immense honour, the honour of power. If we accept the sledded Polacks, this heroic old monarch has to be pictured interrupting an honourable parley to 'smite' his sledge-seated peaceable counterparts. An Englishman, surely, can hardly shoot a sitting bird without indecorum. The image is absurd. Among the many facets of its absurdity, there is nothing to tell the interested why the Polacks didn't get up and smite the Danish King even harder in return. The sledded Polacks in short make the imagination welcome a Shakespeare who writes nonsense—a nonsense underlined by the element of anticlimax in Horatio's image. Horatio's four lines and two images are clearly, firmly shaped and structured so that we look for either contrast or climax (or both). On the one hand, Hamlet combated the ambitious Norway; that he then smote the seated Poles is not another hand, but merely a flat descent in honour.

I suspect that the line's problem is not in fact the pole-axe but the 'sleaded' of the Quarto, emended by the Folio to 'sledded', an image that seems guaranteed to arouse almost Orson-Wellesian nostalgias in Shakespeare editors. It seems possible that 'sleaded' (whatever it meant) held all the explanatory energy of the line in Shakespeare's mind and manuscript. The likelihood is therefore that the word contains or conceals another of great point and energy; and, secondly, that the word in question was difficult or obscure or obsolete enough (like 'ele' and 'overcloud') to prove problematic to the Folio printers more than twenty years after the play was written, these being twenty years of intense linguistic change and development.

'Sleaded' happens to come under some of the same linguistic rules as 'eale' and 'overcloud': we may suppose both the *t/d* confusion already mentioned, and also the 'slide' of *ea* upwards to *e* and downwards to *a*. Since 'slet' is linguistically unlikely, this leaves 'sleat' as a variant of 'slat'. And the OED does indeed offer a verb 'to sleat' as an obsolete version of the verb 'to slat', meaning 'to flap, cast, dash, impel quickly and with some force', a word easily confusable with another, differently derived and now dialectic verb 'to slat', meaning 'to split'. OED illustrates the first verb from Marston's *Malcontent* (1604): '*Men.* How did you kill him? *Mal.* Slatted his brains out'; and the second with two quotations interestingly suggesting the verb's context at the time: '[the nail] slatteth and shivereth in the driving into two parts' (1607), and 'Both head peeces and habergeons were slat and dashed a peeces' (1609).

I am proposing that we ought to take the Quarto's 'sleaded' as only mistakenly emended to 'sledded' by the Folio printers, puzzled by its further move (by 1623) into the archaic; that 'sleaded' was in fact a variant of or literal slip for 'sleated', itself an obsolete form of 'slatted'; and that 'slatted', when set down by Shakespeare as 'sleated' or 'sleaded', so compounded two verbs together as to mean, most vividly, 'smashed' or 'shattered'. Horatio's line has in fact to be radically rethought. King Hamlet, the angry parley cramping his simple features in a violent frustration ('So frownd he') smites the 'slatted' (which is 'shattered') pole-axe on the ice: 'ice', because only winter brings the requisite hardness to the ground-surface, and the requisite brittleness to metal. The phrase has evaded printers and editors just because it is indeed difficult: the words form a packed Elizabethan idiom comparable to the well-known prolepsis of Keats in 'Isabella', 'the two brothers and *their murdered man*'. 'Smote the slatted pole-axe on the ice' in fact means 'Smote, and so shattered—or, So smote, as to shatter—the pole-axe on the ice': the brutal abruptness of the syntax mimics the gesture.

This almost Anglo-Saxon hard density of sense has a peculiar relevance to what Shakespeare is doing in the line. His image is, like its verbal form, self-withdrawn and menacing, very far from that anticlimactic attack on seated Poles. Horatio briefly and tacitly hints at the dead King's haunting ambivalence. The royal Man of Power (his true heir Fortinbras, or 'Strong-Arm'), is a vital defence in a corrupt world: he combats 'the ambitious *Norway*'. But in peaceful, civilized, as it were ice-bound diplomacy, like some mad Hotspur or frustrated Coriolanus, the old Warrior King may simply beat to pieces his own weapon—as the father crying for revenge will helplessly destroy his own son.

I hope that the examination of these two cruces, and of editorial reaction to them, may have suggested how scholarly minutiae can be governed by large critical preconceptions. The reverse is true too: the meaning of *Hamlet* is located in an epithet or stage direction. Thus, the 'dram of eale' and the 'sleaded pollax' may, properly considered, do as much for our sense of Shakespearian identity as might (say) the discovery of *Love's Labours Won* or *Cardenio*, should these interesting fictions prove to have existed. It is at any rate on these grounds and with this preface that I offer the following tentative suggestions concerning other of the problems of this text, numbered for clarity and beginning with (3) to allow for the earlier two.

(3)

After the Ghost's first brief appearance, Horatio introduces (I. i. 80–107) a topic summarized by the biblical 'wars, and the rumours of wars': perhaps thus evoking what might be called the merely public and peripheral causes of haunting, that they may be—as in the later halt called to the advance of young Fortinbras—similarly discarded from the centre of our attention. At all events, this is not the most thrilling moment in the play, and audiences shuffle as Horatio tramps through Denmark's immediate past, telling how the country is arming against the King of Norway's young nephew Fortinbras, who advances in hope to regain by war the colonies lost by his dead father of the same name in a challenge most foolishly and aggressively thrown out against old King Hamlet, but now most legally observed by the Danes:

> *Fortinbrasse* . . . by a seald compact
> Well ratified by lawe and heraldy
> Did forfait (with his life) all these his lands
> Which he stood seaz'd of, to the conquerour.
> Against the which a moitie competent
> Was gaged by our King, which had returne
> To the inheritance of *Fortinbrasse*,
> Had he bin vanquisher; as by the same comart,
> And carriage of the article desseigne,
> His fell to Hamlet.

So the Quarto (I. i. 86–95). The chief of the changes in the Folio is its reading of 'Cov'nant' for Quarto's 'comart', a word which certainly does not occur anywhere else; the Folio word, with its internal comma, perhaps suggests a painstaking attempt to transliterate what its compositors, who worked with Q2 in hand, did not believe to be 'comart'. Modern editors divide between those who, like Jenkins, follow the Folio; and those who, like Edwards, retain Quarto's 'comart'—though not necessarily all for the reason he gives in his edition: the whole line 'suggests a difficulty in Shakespeare's MS which it is useless to try to disentangle'.

Edwards in fact extends his criticism further, finding the 'comart' line and that which follows it 'obscure and repetitive', and suggesting that 'Shakespeare intended to delete all of [the line after 'comart']'. His criticism has the useful side-effect of directing attention to a feature of the passage apparently never noted, its highly idiosyncratic stylistic quality. I have suggested that audiences only half-listen as Horatio drones on, and the nature of this drone needs to be rescued from being called 'obscure' and 'repetitive', since it is something more specific than either. Horatio is plainly pretending to be a legal document, and the dramatist is amusing himself evoking the legal style of the time. In part, Shakespeare is locating in Horatio and Danes like him the possibility of decency—this would have been a Just War, Fair Play. But more: as the speech unwinds Shakespeare brilliantly sketches in the strange domain of 'public life', that extends all the way from this speech on the one hand—with all its sound of what the Army in our own last war learned to call *bumf*, jargonized Civil Service goings-on—to, on the other, the absolute and terrible appearance of the Ghost: who does what the guards cannot do, and stops Horatio.

The Folio printers' doubts of 'comart' are persuasive, in the absence of evidence for it. But, given the force of Horatio's magisterial manner, 'cov'nant' won't do either—its weak trochaic stress is quite wrong for the rhythm of the line, which demands an iamb. A repetition of 'compact' is a possibility. But on consideration I would propose 'contract', and suspect the existence of a legal formula 'contract | And carriage', meaning 'bond and its carrying-out'. The noun 'contract' early carried the second-syllable stress which the line demands and which only the verb has now, a shift which may have masked the word from the beginning.

(4)

The line that follows 'comart' in the Quarto there reads 'And carriage of the article desseigne' (F1: 'Article designe'), a phrase which most modern editors emend to 'article designed' on the grounds of the very frequent *e/d* confusion in secretary hand (i.e. 'designd', as in 'frownd'); only the new New Cambridge leaves the phrase, modernized, as it stands, supposing an intention to delete on Shakespeare's part.

Given the same *e/d* confusion, there is more to be said for shifting the participle and so reading the lines

> as, by the same contract
> And carriage of the articled design,
> His fell to Hamlet

A 'design' is (*OED*) 'a plan or scheme conceived in the mind and intended for subsequent execution' (a meaning nicely echoing 'contract and carriage'). And the design mistakenly hatched by Fortinbras Senior and necessarily accepted by Hamlet Senior was clearly exact, high-level, and legal enough to be 'articled'—past participle of the obsolete transitive form of the verb 'to article' meaning 'to arrange by treaty, or stipulations'; *OED* quotes North's *Plutarch* (1580): 'in which parly it was articled, that the Romans should pay a thousand pound weight of gold'.

(5)

At I. ii. 129 in the Quarto the Prince, left isolated by the departing Court, begins his first soliloquy:

> O that this too too sallied flesh would melt,
> Thaw and resolve it selfe into a dewe . . .

As it happens, the corrupt Q1 supports Q2 here in printing 'sallied'. But the Folio reads 'solid' and is followed by many modern editors, including Edwards; the rest, Jenkins among them, adopt the reading of the earlier New Cambridge, 'sullied', proposed by its editor Dover Wilson with the explanation that Quarto's 'sallied' is merely a regular alternative spelling of the word (as is clearly the case when 'sallies' occurs later in the play).

This is a crux which it is hard to be decisive about, and just possibly the text reflects some uncertainty in the writer.

'Solid' and 'sullied' both make sense, which is more than 'sallied' does, given that the word seems not to exist. Yet I feel a degree of hesitation in believing in a Hamlet who reacts to Claudius by finding himself either 'fat' or 'dirty': neither quite rings true. It may therefore be worth saying something for 'sallied', even if with great tentativeness. The word has already been defended by Furnivall on the ground that it could mean 'assailed'. The trouble with this is that a past participle used as a passive epithet can hardly be so made out of the intransitive verb 'to sally' meaning to sally out in a military sense. There is however a transitive verb 'to sally', which could just conceivably have fused with the other in the writer's mind. This is the obsolete transitive variant of the intransitive verb meaning 'to leap', 'to dance': and it is used of a horse in the sexual sense, 'to leap (a mare)' (*OED*: Sally, v.[1] *Obs. rare.*, 2 *trans.*).

It may be objected that there is an impossible indecorum in making the Prince so describe himself: a reaction which might explain a firm emendation in the Folio from the hand of Shakespeare's printers and possibly Shakespeare himself, as well as of course all succeeding editors. Yet the indecorum is colloquial, now at any rate; a contemporary Hamlet would speak of himself as thoroughly buggered about. Moreover, the Prince has a comparable habit of indecorum throughout the play. His first responses to Claudius in this very scene go far; after the death of Polonius he insists (IV. ii. 54) to Claudius that 'Man and wife is one flesh, so my mother'; and at II. ii. 614–15, in the 'rogue and peasant slave' soliloquy, he comes very near to making the startling 'sallied' allusion again, comparing himself to a 'whore . . . a very drabbe; a stallyon'—all three words, including the last (see (10) below) being female in connotation.

(6)

Hamlet's indecorum is worth stressing in its reaction to the decorum of the world of the Court in which he has his being. Decorum is at once the theme and the style of the scene (I. iii) in which the Polonius family is introduced. And the attempt on the part of the old man himself to achieve a courtly decorum produces at once the moral gaffes in his treatment of his children and, interestingly, the strained and pedantic mannerisms of a style that breeds confusion and the capacity to fox compositors. (In this it should be compared with Horatio's legalisms: the textual difficulty of *Hamlet* can't be dissociated from its virtuoso glitter of rhetorics.) Thus, urging on Laertes a habit of conspicuous expenditure'(I. iii. 73–4), Polonius doesn't quite make sense, either in the Quarto:

> they in Fraunce of the best ranck and station,
> Or of a most select and generous, chiefe in that

or in the Folio:

> they in France of the best ranck and station,
> Are of a most select and generous cheff in that.

I suspect that the problem springs from the old courtier's slightly phoney elegance of vocabulary, an attempt to match the fashions of the French aristocrats he speaks of. He uses 'generous' in a sense now obsolete and probably even in Shakespeare's day rare or archaic, meaning 'of noble lineage, well-born (from the Latin *generosus*)'. The obscurity of the usage surely led the Quarto printers to lose the verb 'to be' that governs it, and the Folio printers to misplace the verb when found. For the sentence ought clearly to say:

> they in France of the best rank and station,
> Or of a most select, are generous, chief in that.

Simply to shift the Folio's 'Are', and to see that it underlay (an easy misreading) the Quarto 'and', is to avoid merely reproducing the Folio line as both Jenkins and Edwards do, and then having to give to the obvious adverb 'chief' (Folio 'cheff') very strained interpretations as a noun ('excellence', 'pre-eminence').

(7)

On Laertes' departure Polonius turns to Ophelia and grows friskier, entangling the two of them in a daisy-chain of happy quibbles; and he ends one such sequence by regarding his own language with the smiling indulgence he hardly bestows on his child:

> tender your selfe more dearely
> Or (not to crack the winde of the poore phrase
> Wrong it thus) you'l tender me a foole.

So the Quarto (I. iii. 107—9); the Folio, perhaps observing that 'Wrong' lacks a syllable, prints 'Roaming it thus'; and is followed in this by Edwards, though most modern editors, including Jenkins, accept Collier's emendation of the Folio, 'Running it thus'.

'Running' has an appropriateness to Polonius's equine image ('crack the winde') that makes it obviously acceptable here. Yet Shakespeare does not always rationalistically sustain the logic of his imagery. And both 'Running' and 'Roaming' are at a distance from 'Wrong' that generates some scepticism. More importantly, neither catches Polonius's point here, the note of good-natured yet maddeningly misunderstanding apology. The old courtier's principles are as awry as his language; his word-games are irrelevant while he treats his daughter's life as a game of power-politics. It is this 'awry' quality which Shakespeare surely located in Polonius's verb. What Polonius accused himself of, ironically, was

> Wrying it thus . . .

and if the poet wrote 'Wriing' the printers' 'Wrong' grows all the more understandable. *OED* classifies as obsolete (but on the evidence of the citations much in evidence in Shakespeare's time, particularly in the context of biblical and theological commentary) two transitive meanings of 'To wry': 'to wrest the meaning of', and 'to pervert'; and both are precisely appropriate, both to Polonius's folly and to his solemnity.

(8)

Polonius ends by defining for Ophelia her own and the Prince's equal unfreedom, even if Hamlet walks 'with a larger tider [i.e. tether]'. His lines (I. iii. 127—31) are a fine expression of the dark inhibition and superficiality of Court love:

> Doe not believe his vowes, for they are brokers
> Not of that die which their investments showe
> But meere implorators of unholy suites
> Breathing like sanctified and pious bonds
> The better to beguide [beguile].

So the Quarto; the Folio corrects 'beguide' but has an error of its own, 'eye' for 'die [dye]'. Polonius's intense but shallow and contorted imagery of money, legal documents, clothes, and ropes or chains makes the lines difficult to rationalize. Malone's emendation of 'bonds' to 'bawds' is now less widely accepted than it once was, though followed by Jenkins—his objection to 'bonds' that 'breathe' being now often countered by the argument that 'breathe' could mean 'talk'. 'Bonds' seems to me justified by 'tider [tether]' and 'springs [springes]'; Polonius's mind turns easily to tying and entrapment.

It might be objected, however, that ropes or chains are as little likely to talk as they are to breathe. There therefore seems to me a possibility that (given the resemblance in secretary hand of some forms of majuscule *W* to some forms of majuscule *B*) Shakespeare wrote

> Wreathing like sanctified and pious bonds
> The better to beguile.

'Wreathing' picks up the images of tying and entrapment by envisaging promises that encircle and coil around the young girl, like ropes of flowers that bind: and the 'wreath' at once unites the pastoral and the fatal, false flowers of love and of death. The image looks forward to Ophelia's mad scene and her reported death scene. It is moreover perhaps recalled in Palamon's dark invocation to Venus (*The Two Noble Kinsmen*, v. i. 95–7):

>   thy yoke
> As 'twere a wreath of roses, yet is heavier
> Than lead itself, stings more than nettles.

Through the Gaoler's Daughter, Ophelia is a haunting presence in this late play.

(9)

A deathly embrace becomes the Ghost's theme too when (I. v. 52–6) he imparts to his son his sense of his wife's depravity:

> But vertue as it never will be mooved,
> Though lewdnesse court it in a shape of heaven
> So but [i.e. lust] though to a radiant Angle [i.e. Angel] linckt,
> Will sort it selfe in a celestiall bed
> And pray on garbage.

The Folio's reading of 'sate' for Quarto's 'sort' is accepted by all modern editors. Yet there are older uses of 'sort', perhaps forgotten by 1623, which better fit the Ghost's point here than 'sate'. OED offers three relevant senses of 'sort': that which means 'suit', and also two reflexive uses implying movement, 'To form sets or groups by some process of combination or separation', and 'To associate or consort with another or others'. It should be noted that the Ghost carefully, even oratorically balances his contrast, matching the vertue unmoved though courted by lewdness against the vice that moves to court garbage. 'Sate' in no way fits this active movement of degeneracy; the decisive 'sort' does. Moreover, there is an obvious appropriateness in these three separatist, consorting and self-suiting meanings to a Queen who for her own wilful pleasure betrays King Hamlet by cleaving to King Claudius. Shakespeare's apparent fusion of the three doesn't make the word's use simple, and as it grew more archaic in these senses printers and editors understandably settled for 'sate'. But the change sacrifices the Ghost's acrid precision.

(10)

Hamlet bitterly accuses himself for his response to the Player's intensity (II. ii. 612–15):

I . . .
Must like a whore unpacke my hart with words,
And fall a cursing like a very drabbe; a stallyon, fie uponn't, foh.

For the Quarto 'stallyon' the Folio reads 'scullion', a reading accepted by most modern editions including the New Arden and the new New Cambridge: T. J. B. Spencer's New Penguin followed Dover Wilson's earlier New Cambridge in retaining 'stallyon' (modernized), though Spencer was surely the more correct in glossing his reading simply 'prostitute' against Dover Wilson's 'male whore'. There is perhaps something to add to the New Penguin's additional note which describes the Folio 'scullion' as 'acceptable'.

Quarto's 'stallyon' was probably in process of rapidly becoming archaic. Giving the meaning of '(female) courtesan' to a word it marks *obsolete*', OED suggests for 'stallion' a derivation shared by both the French 'estalon' (decoy) and English 'stale' (loose woman), and cites: 'Then followed the worshipful Bride . . . but a stale stallion . . . God wot, and an il smelling was she' (1575), and 'a common staliaunt for all that would come' (1584). The drive of Hamlet's rhetoric makes it unlikely to my ear that he means anything but a sequence of three females conceived of as both promiscuous and vociferous—to this image does he, in his misery, vengefully debase both the much-loved Player and his own self. The Folio's 'scullion' may therefore have been brought in to replace a puzzlingly misunderstandable archaism. It may even reflect the more snobbish and class-orientated attitudes of the later Jacobean period. As such, it may *not* be 'acceptable' as coming from a Prince at least up to this point more concerned with morals than status.

(II)

As a character Horatio is a utility. He has human stability only in relation to the Prince; and that relation is, until the last minutes of the play, loyal but detached—he has nothing of Hamlet's slightly desperate lonely warmth. His coolness contributes its own note to the play scene especially. And at one point (III. ii. 92–4) this courtly guardedness produces an elusive meaning which, as in his own earlier legalism or the strained eloquence of Polonius, leads to textual problems. To Hamlet's nervous rapid instruction to watch the approaching King he responds with a soothing

> Well my lord,
> If a steale ought the whilst this play is playing
> And scape detected, I will pay the theft.

F1's emendation of 'detected' to 'detecting' has been universally followed. Yet it strikes me as perhaps a puzzled emendation of a tricky but correct original. The Folio phrase is only at first sight more sensible. 'And scape detecting' is not unlike talk of undiscovered murders in the London area; strictly speaking, if 'undiscovered', murders don't exist, because 'discovery' *is* the imputation of criminality to what would otherwise be interpreted as accident or natural death. If Claudius 'scapes detecting' there will be no theft to pay. The obscure but more exact Q2 phrase in fact compacts some of the moral and philosophical problems of the play. Its 'scape detected' is a diplomatic, difficult and highly Shakespearian ellipse meaning 'escapes with the crime evident but the criminal's responsibility unproven'; the Clowns in the graveyard scene will, perhaps, say something comparable about moral blame for Ophelia's death: 'She drowned her selfe in her own defence'. Because it underlines the fact that even faithful Horatio is capable of cautious clumsy evasions while he is in Court, 'scape detected' ought to be retained.

(12)

While Horatio keeps his eyes on Claudius, Hamlet begins a campaign—partly purposeful, partly helpless—of offensiveness to everyone within range, especially the equally helpless Ophelia. Invited to sit by the Queen, the Prince opens fire with (III. ii. 116)

> No good mother, heere's mettle more attractive.

Both Quarto and Folio spell this reference to Ophelia as 'mettle'; modern editors universally emend to 'metal', presumably guided by the pun on 'attractive'. Although the two words were once one, the different spellings and meanings bifurcated early. While retaining a word-play that asks to be noted, 'mettle' carries meanings lost to us now but I suspect primary to Shakespeare's purpose at that dramatic moment. 'Mettle' implied a vigorous gamesomeness essentially sexual in energies. *OED* suggestively lists first the meaning, 'Of a horse, and occas. of other animals; Natural vigour and ardour', and its citations of the word's use as an adjective (=mettlesome) plainly express lustiness in females. Whatever we think, and whatever Hamlet thinks, of his real meanings here, their aggressive ironies pass into the tragedy and have their sad climax in Ophelia's mad scene. They should therefore be acknowledged by preserving the spelling 'mettle'.

(13)

Ophelia having innocently asked for some practical criticism of the dumb-show, Hamlet obliges with one of the play's darkest cruces (III. ii. 149), which Quarto prints:

> Marry this munching *Mallico*, it meanes mischiefe.

F1's version differs in several respects: 'Marry this is Miching *Malicho*, that meanes Mischeefe.' Modern editors, including Jenkins and Edwards, all print slightly differently punctuated and spelt versions of the *first* half of the Folio sentence, plus the *second* half of the Quarto sentence, (e.g. Jenkins: 'Marry, this is miching malicho. It means mischief'). 'Miching' is explained as an adjective from the present participle of the verb 'to mich' (to pilfer, to skulk, to lurk) common at the time; and '*Malicho*' is said to be an English version of the Spanish noun 'malhecho', or deed of malefaction.

Linguistic scholarship such as the present writer does not possess is needed to make any real contribution here. Yet when a difficult crux has long settled into acceptance there is perhaps room for fresh thinking. It is interesting that, as

J. M. Parker first pointed out, the Folio text of *Coriolanus* spells 'malice' in a form close to the Quarto text here, '*Mallice*' (the italics are in the original text, and may therefore be taken as not necessarily implying personification or foreign derivation in Hamlet's line). If 'malhecho' may be questioned, so also may 'miching' be. Strong as are the arguments for this epithet, the Quarto 'munching' is a curiously bad misreading of what would seem an easy word (though secretary handwriting frequently loses proper count of the indistinguishable minims of letters such as *m, u, n*). It may therefore just be worth recording the existence of a word that links the Quarto 'munching' with the Folio 'miching': a word which, at that time and since, was rare and archaic enough to explain the Folio compositors' failure to recognize it.

The word has, moreover, a Shakespearian connection. In the later 1590s the dramatist had lodgings in Bishopsgate, London, where his local parish church (it still stands) was Great St Helen's. Dealing with the Tower Street area in his *Survey* (1603), Stowe refers to 'Mincheon [now Mincing] lane, so called of tenements there sometime pertaining to the Minchuns or nuns of St Helen's in Bishopsgate Street': and the existence of the lane itself strengthens the possibility of the word being familiar to Shakespeare. By 1658 the term was felt to be so dated as to be referred to as 'Minchings, an ancient word for those consecrated, whom we call Nuns'. 'Minching' can also be used attributively, as in 'minchen clothing', i.e. convent-clothing. Conceivably Shakespeare made his verbally-sophisticated and thoroughly-misbehaving Prince tease the naïve Ophelia with an ironic gloss almost Gothic in its archaism. He describes his Player Queen as embodying 'minching malice': that sweet-faced, demure meanness of spirit which always, as Hamlet says, 'meanes mischiefe'—threatens bad trouble for somebody. The phrase, which is at any rate no odder than 'miching Malicho', becomes part of a quotation, or pseudo-quotation, from some dated alliterative poem both anti-clerical and packed with wise saws: 'Marry, this minching malice, it means mischief.' And it comes with peculiar pain and venom from a Hamlet who has already three times ordered Ophelia, 'To a Nunnery go, and quickly too'.

(14)

The abrupt departure of the King from a Play that seems scandalously to threaten his life leaves a Hamlet violently yet (the text hints) wretchedly excited, a mood reflected in the way the Quarto presents the speech climaxing his rebuffs to the attempts by Rosencrantz and Guildenstern to corner him: 'Why looke you now how unwoorthy a thing you make of me, you would play upon mee . . .'. To this passage (III. ii. 379–90), Q2's extremely light punctuation gives a rapid and even hectic movement, culminating in Hamlet's snapped dismissal:

> call mee what instrument you wil, though you fret me not, you cannot play upon me. God blesse you sir.
>
> *Enter Polonius*

The Prince is playing with a double sense of 'fret' here, meaning both 'to irritate' and 'to furnish (a guitar, etc.) with frets'—i.e. with 'a bar or ridge of wood, metal, etc. placed on the fingerboard, to regulate the fingering'. The Quarto's 'fret me *not*' seems surprisingly to negate the very word-play involved. Modern editions therefore make alterations. All emend, most to the Folio's 'can fret me':

> Call me what Instrument you will, though you can fret me, you cannot play upon me. God blesse you Sir.
>
> *Enter Polonius*

In the New Arden, Jenkins emends to 'fret me'. In addition, all modern editions except the New Penguin move 'God bless you sir' after the stage direction, thus making it a greeting to Polonius rather than any address to Guildenstern.

In supposing the Quarto printers to have simply negated the plain meaning of their text by inserting a whole word we see them as capable of thoroughly bad work. Is there any other explanation? The Prince himself is, at least according to his own Renaissance code of honour, behaving discreditably here, lacking *noblesse oblige*, all-but-shouting at two Court servants who were sometime friends of his own. The sudden self-consciousness of the 'fret me' phrase is surely an acknowledgement of this loss of self-respect, part apology and part further rage. The muffled desperation here suggests that Hamlet may originally have echoed his opening 'Why looke you now . . .' with a stressed repeating 'you fret me now'. The 'now' will have been misread by the Q2 printers for the simple reason that they were as surprised by this lack of detachment as the Prince himself and so read in the calm 'you fret me not' which they expected, failing in the process to take account of the spoiled pun.

If we accept 'you fret me now' with all it communicates, then the abrupt 'God blesse you sir' is clearly a would-be courteous dismissal of Guildenstern, who has provoked all this. As such it makes more sense than as an oddly-phrased welcome to Polonius. We should therefore retain the Q2 text, while emending its 'not' to 'now'.

(15)

Hamlet's violent agitation is mirrored in the King's business-like decision to get rid of him (III. iii. 1–7). He instructs Rosencrantz and Guildenstern, who have promptly reported back to him:

> I like him not, nor stands it safe with us
> To let his madnes range . . .
> The termes of our estate may not endure
> Hazerd so neer's as doth hourely grow
> Out of his browes.

Though most recent editors including Jenkins and Edwards retain a modernized form of Q2, 'browes' has generated discussion and emendation, a debate taking into account the fact that F1 seems to reveal its own doubt of the Quarto text; it makes radical changes in the last two lines here and reads

> Hazard so dangerous as doth hourely grow
> Out of his Lunacies.

It is probably a mistake to treat both 'dangerous' and 'Lunacies' as identical cases, as most modern editors do, assuming that 'browes' foxed the compositors of F1 in exactly the same way as 'neer's', and that they were as wrong to replace it. 'Neer's and 'browes' may have seemed to the F1 compositors to be both archaically senseless; but the fact remains that 'near us' makes sense now, and 'browes' does not—though Jenkins explains it as 'plots, contrivances' and Edwards as 'effrontery'.

To anyone unconvinced by these glosses the dramatic context offers information that has, I think, hitherto gone unnoted. The King's language to the two courtiers is stylistically conditioned in the ways earlier proposed for both Horatio and Polonius. He is entrapping Hamlet in words of wary political power, diminishing even as he denigrates. The King broadcasts an image of himself as superbly calm even while threatened: the Prince's actions are mere 'madnes', are as random as 'Hazerd'. For this last is a word which did not in Shakespeare's time mean merely 'danger' (as we may and the Folio printers probably did assume) but something nearer to 'playful randomness, triviality': 'Hazard' was for Elizabethans the name of a game of dice. Claudius is seeking a coolness close to that of the schoolmaster in the old joke who, finding his classroom in flames and half his colleagues butchered, remarks that some dangerous buffoon has been at work: a tone beautifully matched when at the beginning of III. iv Polonius passes on to Gertrude what we feel at once to be the King's message:

> Tell him his prancks have been too braod [*sic*] to beare with.

'Browes', which the Folio printers rightly if fruitlessly tried to correct, conceals beneath itself a word that puzzled the Quarto printers equally; it will therefore prove to be difficult, obscure, or archaic enough to explain both sets of confusion; and it must mean something closely in harmony with 'madness', 'hazerd', and 'prancks too braod [*sic*] to beare with'. I believe that this word, one easily mistaken for 'browes' in secretary hand, was the now obsolete 'bourds' (found in many other spellings, e.g. 'bowerds'). OED defines 'bourd' as 'an idle tale, a jest, a joke; jesting, raillery, joking, merriment, fun; a merry tale . . . mockery, bantering . . . play, game'; and it quotes relevantly, 'If a man . . . should strike another or use broad boward against him . . .' (1602). In Claudius's Court the struggle against evil will justly or unjustly be played out as 'broad bowards', and their 'Hazerd . . . neer's'.

(16)

The pervasive presence of 'bourds' in *Hamlet* is reflected in the casting of the Clown as Gravedigger. And the Gravedigger lets fall a phrase which usefully demonstrates how easy it can be to lose a sense of an art-work's proper realism. Dismissing his assistant (v. i. 67) to leave himself free for Hamlet's encounter, the Clown says what the Quarto prints as

Goe get thee in, and fetch mee a soope of liquer.

F1 emends and elaborates:

go, get thee to *Yaughan*, fetch me a stoupe of Liquor.

Yaughan, explained as a Welsh pronunciation of Johann, has had an interestingly vivid existence among editors. Only the New Penguin retains the Quarto reading; all other modern editions including Jenkins and Edwards follow the Folio, telling us that Yaughan was (as Johann) a London alehouse-keeper. Nobody explains, however, how Johann got to Wales. More seriously, no one finds reasons why Shakespeare, with so evidently little gained by it at this point of his tragedy, should have blurred his effect by the introduction of some quasi-Jonsonian contemporary allusion.

Significantly, the Quarto's 'in' is odd too: 'in' where? The chances surely are that *to Yaughan* and *in, and* are equally attempts to read a third and illegible phrase (there is a suggestive clue in the comparable endings, 'an' and 'and'). I pointed out above (p. 141) that, as in 'French/friendly Faulconers', the arch and descender of some forms of secretary *h* could deceptively suggest the presence of a *y*. To this must be added the fact that *y* was the standard graph which had replaced OE *þ* as a symbol for *th* in such words as 'the', 'this', etc., and had thus produced the non-existent word '*ye*'. We may explain the *y* of 'to Yaughan' in the same way. I would conjecture beneath this quasi-name the words

> Goe get the[e] to th'ynn and fetch me a stoup of liquor.

In the phrase here underlined, in secretary hand 'the[e]' and 'th'' would both contain a 'y', of which pair one would be likely to be dropped through haplography, and the other 'y' has caused confusion. Quarto compositors may have simplified to the only legible words, 'in, and'; the surprised Folio printers would have looked again at the manuscript and conjured up the Welsh-speaking German taverner Yaughan. A modern text should read:

> Go, get thee to the inn and fetch me a stoup of liquor.

(17)

After the graveyard scene the tragedy's element of 'bourd' or savage game comes to its climax in the duel, to which the Prince is called by one whom the Quarto names (v. ii. 80) in its stage direction, 'a Courtier'. This 'Courtier' is at once succeeded by another summoning figure, 'a Lord', who refers to his predecessor as 'young Ostricke': a form of the name retained by Claudius shortly after, at line 270. Later again, however, when this courtier enters to announce the arrival of Fortinbras, the stage direction—correctly or incorrectly—calls him 'Osrick', with a speech-heading '*Osr.*'. The Folio throughout names him '*Osricke*'. Making some cuts in the earlier part of v. ii, it excises the 'Lord', digesting his lines into the part of the Courtier, firmly named on arrival in his stage direction (v. ii. 80) as '*Enter young Osricke*'.

All modern editions use the Folio form of the name. This is a form which it is easy to feel that actors have probably always found more sayable (even apart from the fact that the name 'sounds Danish'). Yet the New Arden raises a query interesting enough to make it worth while suggesting that

Q2 deserves some attention. Jenkins remarks that Osric 'is a notable Anglo-Saxon name, and as such occurs in earlier Elizabethan drama . . . It is not apparent why Shakespeare gives it . . . to one who inspires contempt'. If 'Osric' seems wrong both in terms of nationality and of moral quality, then perhaps 'Ostricke' deserves a fairer hearing.

At v. i. 7 in the Folio version (the only original text) of *All's Well That Ends Well*, a stage direction reads '*Enter a gentle Astringer*'. Though invariably emended to 'Enter a Gentleman' or 'Enter a Gentleman, a stranger (i.e. one who has not appeared before)', this should I believe take the form preserved in my own (New Penguin) edition of the play: 'Enter a Gentleman Astringer'. In this form it gives an important indication of Shakespeare's intention to portray a courtly milieu. An astringer is a keeper of goshawks, an essentially courtly occupation. The word itself is found in many forms ('austringer', 'ostringer', 'ostriger') all deriving from the Middle English 'Ostreger', which in its turn seems to be a corruption of Old French 'austruchier'. Our own modern 'ostrich' is similarly descended, but should not (I believe) be allowed to confuse the issue of the Courtier's name.

For 'Ostricke' was surely conceived as a more mannered first cousin of the gentle Astringer of *All's Well*, or vice versa: both are courtly hangers-on attached to the royal Mews. The professional connection gives a certain mean point to Hamlet's aside to Horatio involving a 'chough', a bird of the crow family, possibly a jackdaw, and hence a chatterer (though there may be further word-play on 'chuff' meaning 'bumpkin'). Whether a modern editor ought to call the character 'Ostricke', 'Ostric', or 'Osric' is a difficult question. At any rate, it seems a pity that the New Arden, for instance (probably the best and fullest edition of the play that we now have, and the most widely used) nowhere gives any hint, in notes or textual apparatus, that Quarto's reading of the name might be other than error. The New Penguin *Hamlet* of T. J. B. Spencer (sometimes perhaps textually the more sure-footed of the two) at least raises the possibility even though finally rejecting it. I believe that the 'austruchier/astringer' background of the name does fill in, with curious vividness, some small detail of that Court world Shakespeare imagined for his tragedy. 'Ostricke' does something that all such new readings may do: gives us more sense of the real *Hamlet*, and therefore the real Shakespeare.

8
The Inaction of *Troilus and Cressida*

TROILUS AND CRESSIDA has no story, or is as near to having none as a Renaissance play can be. And it is the only one of Shakespeare's plays of which this can be said: no other, for instance, shares its notable lack of a formally structured ending, as of wedding or funeral. This absence of simple story-line in *Troilus and Cressida* is experienced by anyone who tries to remember the exact order of events, particularly in the middle of the play; and it is the source of most of the other problems that disturb the play's readers. No one finds it easy to understand how the play's action develops, if indeed it does develop; or to decide who its chief characters—the protagonists—actually are, as between Troilus and Cressida themselves, who hardly come in, Ulysses, who is a bore, Achilles, who is a thug, Hector, who is opaque, even Thersites, who does nothing; and all of them have decidedly little to do with one another. Since what characters have to do with one another is a basic part of the meaning of 'story', this is what in the present case appears to be missing.

This strangeness in the play is all the more remarkable in that it uses the materials of two of the greatest stories known to the writer and his contemporaries. The infidelity of Cressida forms the climax of what was probably the greatest medieval poem available to Shakespeare, who did not read Dante; and though Chaucer's story took on changes in its transmission down through Henryson, with whose poem it was bound up in the version all Elizabethans read, it still kept there the sense of great story: the climax—here, Troilus's only-half-recognition—remained the point. Similarly, most readers of the time, including Shakespeare, may have been more familiar with the matter of Troy in Caxton or in Lydgate than in the Homeric original, since when *Troilus and Cressida* was written Chapman had completed only the first seven books of his *Iliad*; but from *some* translation or redaction most readers of the age knew the story that had as its great climax in Homer's penultimate book the death of Hector, with the last and twenty-fourth going on to tell how an Achilles at last moved to pity gave up to a grieving Priam the body of his dead son.

We make contact with these great stories at two points in Shakespeare's play. When Hector is killed, Troilus comes to the Trojans and tells them

> *Hector* is dead: there is no more to say.[1]
>
> (v. x. 25)

After Cressida has betrayed Troilus, Ulysses begins to draw him away from the scene:

> Al's done my Lord.
> It is.
> Why stay we then?
>
> (v. ii. 135–7)

It is interesting that Troilus's utterance after the death of Hector makes use of a characteristic Chaucerian phrase, as if quotation were also denotation—a gesture towards some original story. At these two moments, with these pointers to the fact that something is finished, that something and someone has died, which is the first evidence of completion in the play (as though death at least were an inviolable structural principle), we perceive that in Aristotelian terms something has taken place, something is *done*. There has been action and it is, surprisingly, the action of the great stories: surprisingly, for the action has not been played out in terms of sequential logic on the stage before us, but only in terms of our own conscious expectation, our own simple historical sense. For two things we surely know from the beginning about these ancient narrative materials: that Hector must die and that Cressida must be false. Whether 'history' is the epic

---

[1] Throughout this essay I quote from the New Variorum text of *Troilus and Cressida*, which is based on the Folio. Some readings therefore will be unfamiliar: e.g. 'Distasting' at IV. iv. 47.

and classical history of the Greeks, or the romantic and idealizing history of the Middle Ages, or perhaps even our own more inward sense of what has been and therefore will be, these are historical data so absolute that they cannot be omitted or distorted or displaced. Therefore unexpectedly the expected has taken place.

What is peculiar about *Troilus and Cressida* is the degree to which the expressive self-containment of the old stories has been replaced by this activity of a quasi-modern 'consciousness': the way in which this expectancy in us has to *constitute* narrative. *Troilus and Cressida* is a play because we *know* it is a play (hence the advertency and intimacy of Prologue and Epilogue); it makes us rejoin our shared past at the instant Hector is killed and Cressida is false, at the instant, moreover, that we know that we know these things. Any more conventional sense of story has its back broken very early on, by the debates in the Greek and Trojan camps; and these debates get something of their grotesque supererogatory bulk, their fascinating boringness, from the fact that we *know*, and the characters know that we know, that the Greeks did not really have to worry why they were not winning the war, because they were going to win the war; and the Trojans did not have to debate whether to send Helen back, because Helen had come to stay. And it is the inbuilding of this knowledge which produces the harsh farce of the Greek style and the slightly second-rate nobility of the Trojan—in neither case quite 'satirically' handled, but in both decidedly knowingly; as it creates, too, the decorum of each grave debate, one of policy, the other of value. Each party has in its own way to know what it is doing, although none of the arguments amounts to the supreme blank knowledge that we have.

From the Greek council comes as solution to its problems the intrigue of Ulysses. This bulks large in most attempts to tell the story of the play. But its presence there, in fact, seems to say only that whatever real action is, and whatever real consciousness, they are not this cerebral political game. For the plot of Ulysses is notoriously pure 'plot'—plot in inverted commas: it comes to nothing; from the beginning it comes to nothing. Its coming to nothing in a technical sense is a thing that we should have expected as much as we should have expected Hector and Troilus to come to nothing, humanly speaking. That is to say, the reasons for the failure of Ulysses's stratagem to incite Achilles to action by arousing envy and anxious ambition in him, are not accidents of narrative, here wantonly random (Hecuba playing *dea ex machina*) but are innate in the formulation and conditions of the plot itself. This is proposed in answer to Agamemnon's and Nestor's problems. Though each of these two has, like everyone else in this play, only a sharply delimited amount of 'character', each does here have an identifying style, and it is a style of empty power. Each 'ample proposition' by the powerful Agamemnon 'fails in its promised largeness'; each epic simile launched by the fragile pointless old Nestor floats away on a sea of irrelevancy among 'shallow bauble boats'. Agamemnon therefore turns to Ulysses, and this is his way of saying '*You* try, Ithaca':

> be't of lesse expect:
> That matter needlesse of importlesse burthen
> Diuide thy lips; then we are confident
> When rank *Thersites* opes his Masticke iawes,
> We shall heare Musicke, Wit, and Oracle.
>
> (I. iii. 76–80)

Negation, however circumlocutory, could hardly be more triumphantly self-confounding. It is to the point that when Aeneas arrives a little later as ambassador his similarly self-defeating though superlative politeness leaves Agamemnon baffled as to whether he is being funny or not. In this milieu it is not surprising that when the evidently more 'brainy' Ulysses—appealed to *for* his braininess—intervenes with a remarkably superior clarity, his braininess is not in the end all that intelligent, either. We feel with some part of our minds, by now fully awakened to irony and abraded by the harsh farcical overtones of what passes, that compared to the Greatness of these Great Men, human beings must be small;

THE INACTION OF *TROILUS AND CRESSIDA*

Great Rhetoric therefore will not and cannot serve their needs or propose their remedies. The splendid authoritarian structure of Ulysses's speech therefore in its turn converts into a pure linguistic overreaching, that

> . . . by a pace goes backward in a purpose
> It hath to climbe.
>
> (I. iii. 135-6)

There is obvious distinction in Ulysses's diagnosis of human history—more specifically, of the reasons why they have not yet killed all the Trojans, why (so to speak) Hector is not dead nor Cressida false *yet*: but the distinction is the detachment; and such subjects do not leave us wholly convinced that we *can* be detached from them, or can be so rightly. And the doubt gives an odd hollow ring to Ulysses's prescription of Order, Order, and then more Order. All Ulysses can offer (like Macbeth's doctor failing to minister to a mind diseased) is an orderly analysis of disorder, which sounds paradoxical anyway.

Paradox is what Ulysses accuses Achilles and Patroclus of making: and this climax of his accusation is an extraordinary moment that crystallizes all the storylessness of this inactive play. Ulysses argues that they are not winning the war, that history is being held up, because some of the Greeks are not taking themselves seriously enough. In a small tent Patroclus is play-acting, and

> The large *Achilles* (on his prest-bed lolling)
> From his deepe Chest, laughs out a lowd applause,
> Cries excellent, 'tis *Agamemnon* just.
> Now play me *Nestor* . . .
>
> (I. iii. 169-72)

This is one of the most startling moments in Elizabethan drama: (another is the arrival of the ghost in *Hamlet*, which works in rather the same way). Our surrogates, the actors, are playing Greeks—or merry Greeks, some Elizabethans would have said, meaning playing the fool; and these Greeks played by actors are playing at actors; and these actors played at by Greeks who are really in fact actors suddenly—we being of course speakers with Ulysses and listeners with Agamemnon and Nestor, the named Nestor—suddenly *look at us*, right out of the play, as if it turned into a mirror. This moment of pure self-reflection and self-consciousness stops the play, as it were brings down the house. It is like meeting the eyes of the Beast in the Jungle. Under the light of the instant, Achilles, hardly characterized elsewhere (except maybe as a prize-fighter), turns into Hamlet. Technically the dramatic moment is at the very obverse from the impersonal expressive symbolism of story, to which this scene-stopping self-consciousness is, as Ulysses says, 'done, as neere as the extreamest ends of paralels'. And tellingly, Ulysses proposes a remedy as simply self-conscious as the malady: that Achilles should be triggered into action by vanity, by a concern for his own 'image'. We know intrinsically that nothing real or permanent can come out of this mirror-world of theorizing, of base and imitation self-awareness. And indeed Achilles comes alive only when the mirror is smashed. It is the death of Patroclus that brings him into action, and by that time the play is all but over—over almost before it has seemed to begin.

This peculiar rhythm of late beginning and premature ending may be recognizable primarily from early modern works of literature—James's stories, say, and Eliot's poems. At the end of Kafka's novel *The Trial* his hero K is marched off to die on a waste ground reflecting 'Am I to leave this world as a man who shies away from all conclusions? Are people to say of me after I am gone that at the beginning of my case I wanted it to finish, and at the end of it wanted it to begin again?' Kafka, who left the novel unfinished when he died, wanted everything *before* this last chapter to be as it were infinitely unfinished, incessantly long: but the point where he left it, at least, takes K into a cathedral where he is told by a priest the story of the 'man from the country' who wants to enter within the Law, but waits all his life by a doorway whose doorkeeper tells him only that he cannot enter 'at this moment'. At the end of the book we may feel that the one moment he was perhaps capable of entering within the Law was the moment at which moments ceased for him to exist—the moment of death; but of this K, the ever-conscious and quasi-first-person narrator, cannot ever while alive be conscious. This self-indicting structure has perhaps an earlier parallel, though in a wholly different dimension and mode and tone, in *Troilus and Cressida*. The intrigue of Ulysses could not have worked, even had it not been for the intervention of Hecuba, because it is not 'within the Law'—not, as Elizabethans would rather have said, 'in Nature', not true, merely a mess of shadows and vanities. Ulysses's theory of success is merely a matter of *moments*, it 'shies away from all conclusions'. But his intrigue none the less (like K) serves a purpose of self-indictment: the way it does *not* work has its own meaning, it is a plot that 'proves itself' at the moment at which it betrays its own inutility, its final nullity.

Troilus and Cressida is a war-play that starts rightly with Troilus's 'Ile unarme againe'. As such, not only the intrigue of Ulysses, but all its action has something of this same self-defeating bearing, the self-undermining movement of Socratic irony, which is why—despite its harsh sad content—the play appears to approximate to satiric moods. Probably few artists have ever been more endowed with pity than Shakespeare was; but in this work 'the pity of war' and of love is only equal to an enormously astringent satiric play of mind. Under the force of this acerbic, vigorous, and ultimately exhilarating critique, even the death of Hector and the falling-away of Cressida are subsumed into a self-indicting action. Each of the two great stories becomes an example of the way action or 'doing' is a straight road to *un*doing. The words 'doing' and 'done' reverberate blankly throughout, reminding us that they had for Elizabethans the same cant obscene sense as they still retain; so that the 'All's done' of Ulysses, like Troilus's matching 'There is no more to say', confers something of that brute reductiveness on the actions these phrases describe. The relation of Troilus and Cressida is a 'doing' that undoes the two of them: just as Cressida's

'Things won are done' forewarns that nothing is won by doing, that winning is only undoing. Swift said that 'We have just enough religion to make us hate one another'; Troilus and Cressida have just enough love between them to make up a betrayal. By a time-stopping brevity Shakespeare simplifies Chaucer's lovely slow natural coming-together and slow painful breaking-apart, more than three years together, into a 'one-night cheap hotel', love-making just enough to produce the conditions for severation. More unlike from the beginning than any other two lovers in Shakespeare, Troilus and Cressida are joined by an attraction that is tender, anxious, and pricklingly hostile, the more erotic for its element of opposition:

> He will weepe you an 'twere a man borne in Aprill.
> And Ile spring up in his teares, an 'twere a
> nettle against May . . .
>
> (I. ii. 173-6)

Their love, which contains intensity, is a good way of finding out that they probably did not like and certainly did not know each other: what they do know, whether or not they like it, is that something is finished: ' 'Twas one that lou'd me better then you will. | But now you haue it, take it.' The play's action for them is the substantiation of this knowledge, not really any alteration of their relationship; for nothing surprises Cressida in the play, though she explains what happens after her own fashion: 'You men will never tarry.' And Troilus for his part is already calling her 'Traitor' in the thirty-fifth line of the play, for not always occupying his thoughts. No wonder he sometimes needs to forget her, for she is to him the turn of the screw, the last twist of the knife, an occasion for explaining to himself and making beautiful an inward intrinsic despair that has little to do with her: 'There my Hopes lye drown'd.'

If, under the analytical story-destroying wit of the play, Troilus and Cressida are hostile lovers, for whom love is a way of proving human separateness, the Greeks and Trojans at large are lover-like warriors who find out, in the

THE INACTION OF *TROILUS AND CRESSIDA*

extraordinary mêlée of the close, how horribly little difference there is between body and body. Hector refuses to continue to fight Ajax because his enemy is in fact a blood relative of his own; and Achilles states that he feels

> a womans longing,
> An appetite that I am sicke withall,
> To see great *Hector* in his weedes of peace;
> To talke with him, and to behold his visage,
> Euen to my full of view . . .
>
> (III. iii. 248–52)

In the night-scene of fraternization, that gives an impassive echo to Cressida's comparable willingness, in the event, to adapt:

> To morrow do I meete thee fell as death,
> To night, all Friends
>
> (IV. v. 297–8)

—in this night-scene Achilles makes of the encounter at last with Hector a lethal eroticism:

> Tell me you Heavens, in which part of his body
> Shall I destroy him? Whether there, or there, or there
>
> (IV. v. 267–8)

Blood-brotherhood is one thing (so Hector might rightly feel), being publicly affronted another. And the final murder of the unarmed, as it were naked Hector by Achilles' gang— not even by Achilles himself—brings a conclusion the more horrible in the knowledge of possibility it has violated. If Shakespeare had not invented some relationship between these two great names, some sick half-love like the actual lovers' failed love, there could not have existed that pure experience of betrayal which the death of Hector invokes.

'Betrayal' is the important word here, because it is the governing concept of the play. The two scenes which summarize all Shakespeare's relation to his stories, the scene of Cressida's fall and that of Hector's death, also summarize the dramatist's departure from his stories, and his acts of essential interpretation and revision. We see something of that change by noting that the real 'last twist of the knife' in Hector's case is the unimportance of his death scene: how little of a true climax, how easily missed the death now is. Exactly similarly, Troilus is now not even left alone in his miserable witnessing of Cressida's change of social loyalties: Diomedes interposes, Ulysses stage-manages, Thersites sits on the side-lines, and in front of it all, we stand. This must surely be something of what it means to look 'on truth | Askance and strangely'; for art could hardly be more complicit in selling 'cheap what is most dear'. The very techniques underline the fact that Shakespeare has assimilated his two stories to one another, and then re-created them both—in style as in substance—as a great system of betrayal. The medieval Criseyde was false, but Chaucer's manner is too kind and Henryson's gods too unkind to make her inaction quite betrayal. But Shakespeare has procured an equivalence of opposites. His Troilus is a betrayee, as Lawrence defines certain people as murderees: but he is fully complicit with his own fate. He exists to distrust, with anguish and extremity and doubt and sheer self-will, the mere untrustworthy love-object that Cressida is created to exist as. Where Troilus tends towards pure, inward, formless, messy feeling, pure unsatisfactory subjectivity, Cressida tends to be just (as we say) a pretty face. *All* her expressed motives seem equally assumed and surprise us as though a cat should speak (and speak sometimes with great good sense): this is why her infidelity leaves us *un*surprised, since constancy is not to be expected from the motiveless. This does not mean that Ulysses is right about her: his nonsense about her foot speaking, which nothing in the text validates, is as violently inappropriate as Troilus's nonsense about her soft hand. Both merely apply to her, as if she *were* a cat or a nettle, an external rhetoric which equally reduces her to a thing. But Shakespeare's imagination reduced her first, giving her as much right as Parolles to say 'Simply the thing I am shall make me live'. And her pitiful relationship with

Troilus earns them both the sad epitaph of the courtiers in *All's Well*:

> As we are ourselves, what things are we!
> —Merely our own traitors.
>
> (*AWEW*, IV. iii. 18–20)

Matchingly, Shakespeare has invented a fate for Hector sharply different from all the sources, including Lydgate whom it most resembles. The dramatist sets the two warriors Hector and Achilles in something closely enough approximating to a personal relationship to make of the death an act of pure betrayal. Achilles, betraying his own latent and incipient (though unrequited) feeling for the other man, at the same time by his vengeful thuggery betrays Hector's war-idealism, which has led him immediately before to spare the unarmed Greek. Hector, in his turn, contributes to that process—like Troilus he is complicit with his own betrayal—by his dramatically marked collapse at the end of the Trojan council: he betrays his own rational and bonding sense of what is to be counted valuable in the great human order, so as not to be divided from the nearer but also more merely social bonding to his brothers:

> My spritely brethren, I propend to you
> In resolution to keepe *Helen* still;
> For 'tis a cause that hath no meane dependence,
> Upon our joint and several dignities.
>
> (II. ii. 197–200)

'Joint and several', together and divided—the phrase describes the whole self-betraying condition both of the individual mind and of human relationship in this play; where the suddenly decisive Hector and the suddenly 'troubled' Achilles betray themselves and each other, no less than do the lovers, the reluctant, cool little Cressida and the unhappily passionate Troilus, who knows his divided bonds from the very first lines of the play:

> Why should I warre without the wals of Troy
> That find such cruell Battell here within?

* * * * *

I have been hoping to suggest some of the reasons why *Troilus and Cressida* may strike us as a play without a story. 'Story' is not merely a random effect, but presupposes a creative wish to show how things hold together with a certain good purposiveness—a certain true coherence. In a now well-known essay on Nicholas Leskov (written in 1936), the critic Walter Benjamin described what seemed to him to be the decline of the art of story-telling as a great human catastrophe, as well as a distinctively modern experience: 'It is as if something that seemed inalienable to us, the securest among our possessions, were taken from us: the ability to exchange experiences.' Benjamin finds the chief cause of this decline to be a general devaluation of human experience as such, and a disbelief in such wisdom as can be derived from experience: an end, in short, of such thinking as is based on human history. And he traces this in its turn to the First World War, from which

> Men returned . . . grown silent—not richer but poorer in communicable experience . . . A generation that had gone to school on a horse-drawn streetcar now stood under the open sky in a countryside in which nothing remained unchanged but the clouds, and beneath these clouds, in a field of force of destructive torrents and explosions, was the tiny, fragile human body.[2]

There is no reason to think that Benjamin had Shakespeare's play in any sense in his mind here. The foreshadowing, in *Troilus and Cressida*, of certain distinctive elements of modern experience may merely make Benjamin's penetrating remarks helpful in explaining something of the play's strangeness as narrative. If it lacks 'story', it also envisions a world in which experience itself endures devaluation; and in which human beings do not find it easy to 'exchange experiences'. What is remarkable about *Troilus and Cressida* is the degree to which it can reflect so much of the fragmentation and atomism that we think peculiar to modern life, and yet that, lacking

---

[2] From 'The Storyteller: Reflections on the Works of Nicholai Leskov', in *Illuminations*, by Walter Benjamin, ed. Hannah Arendt, tr. Harry Zohn (1973), 83–4.

'story', the play still makes such intense sense of its lack—still finds a design that is in itself an 'exchange of experience'.

For the play does not absolutely lose by its lack of story and its consequent diminution of characters—the lovers hardly more than private shadows of the greater public betrayals, the politicians hardly more than brilliant revue-sketches. The play is distinctively a two-dimensional work, lacking atmosphere just as it lacks story and character, lacking foreground and background as its storyless action lacks past and future; but that two-dimensional area is so thoroughly mastered and understood as to be peculiarly, painfully consistent with itself at every random point on it. Whichever way we approach it, whatever name we give to what we are analysing, every accidental detail will render up an equal betrayal. Thus, the work affords many moments which are, as it were, well out on the periphery of attention, and which cannot be brought within it by any naming of 'story' or 'characters' or even the utilitarian 'theme', and yet which will bring one into the full *Troilus* effect, as though this random moment were the essential crux of the dramatic action. Consider for instance those casual lines with which Ulysses announces chorically that the curtain is rising, the action at long last beginning—ironically, since we have reached the middle of the fifth act of the play:

> Oh courage, courage Princes: great *Achilles*
> Is arming, weeping, cursing, vowing vengeance;
> *Patroclus* wounds have rouz'd his drowzie blood,
> Together with his mangled *Myrmidons*,
> That noselesse, handlesse, hackt and chipt come to him;
> Crying on *Hector*.
>
> (v. v. 34–9)

The brilliantly acid style of *Troilus and Cressida* can encompass a poetry of total war as direct as the images of Benjamin's prose, even though the poet had not himself experienced any 'war to end wars'; and this is perhaps because it makes private and public worlds freely 'exchange experiences'. The passages from Caxton and Lydgate which presumably served Shakespeare for source here are revelatory in the plainness with which they communicate the fact that many Myrmidons were hurt and many killed.[3] An extraordinarily vivid and eccentric wit by contrast animates these

[3] Caxton writes: 'When the battell was finished, against the euen the Mirmidones returned vnto the Tent of Achilles, & there was founde many of them hurt, and there were an hundred of them dead, whereof Achilles had much sorrowe: and when it was night, he went to bed, and there he had many thoughts, and purposed once to go to the battell for to reuenge the death of his men . . .' (*The Auncient History of the Destruction of Troy*, Book 3, ch. 22; see New Variorum *Troilus and Cressida*, 445.)
Lydgate writes:

> And to Achylles lying in his tent,
> They be repeyred forwounded and to rent.
> Their herneys broke both in plate and maile
> And of their nombre I fynde that they fayle,
> A hundred knyghtes slayne dead alas . . .
> . . . .
> At his herte his wounde was so kene,
> What for his men and fayre Polycene.
> Wyttinge well if he dyd his payne,
> To be venged he shulde not attayne,
> In no wise vnto his desyre.
> And thus he brent in a double fyre,
> Of loue and yre that made him syghe sore . . .

(*The Auncient Historie and onely trewe and syncere Chronicle of the Warres*, ed. Bergen, Book 4, ll. 2337 ff.; see New Variorum, 445.)

Something of the difference between Caxton and Lydgate, and Shakespeare's account of the Myrmidons may be explained—or so I would suggest—by the intense impression made on the dramatist by two or three words from Aeneas's account of the Fall of Troy in Marlowe's *Tragedy of Dido, Queen of Carthage*. This speaks of the '*mangled* head of Priam's youngest son', and of the 'band of Myrmidons. | With balls of wildfire in their murdering *paws*'. Shakespeare has remembered the 'mangling' but has, with an even-handed poetic justice, returned it back on to the Myrmidons: whom he has been helped to create by Marlowe's brilliant image of flame-throwers in the paws of animals. Marlowe, of course, like most Elizabethans (but unlike Shakespeare) shows no sign of sympathy anywhere with the animal world; therefore the image is in Shakespeare acutely complicated by his empathizing incapacity to withhold a kind of pity even from the Myrmidons, once seen as animals. Furthermore, it seems possible that Shakespeare then imagined Ajax as very like the Myrmidons, and gave him this same ambiguous and grotesque animality, from which sympathy or at least humour is not totally withheld ('Hee's growne a very land-fish, languagelesse, a monster', III. iii. 273–4); and, further again, that the creation of Caliban later on may have profited by this rich compound of ideas.

lines from Shakespeare's play, which bring the massive armoured fighters on with a heartless derision, a long-distance historical objectivity that turns them into—precisely—objects, *things* to be absurdly 'mangled', 'hackt and chipt'. But at the same time, paradoxically, this wit is transfused with an ambiguous pity. The Elizabethan phrase 'Crying on' meant 'exclaiming against', but the words leave room, because of the blankness of the preposition and the plangency of the verb, for a more modern reading, for a feeling far more childish and brutishly inarticulate, as of small bloodied beaten-up boys saying 'Hector did it', or a yowling animal holding up a broken paw. This dry vision of the mangled Myrmidons, half painful sympathy and half savage humour, is supported by what precedes it, the historic image of 'great Achilles', in the first line locked up in a heroic brazen rhetoric, but in the second collapsing into a pure extremity of human tears and rage, 'arming, weeping, cursing, vowing vengeance'—mastered by time instead of obstinately mastering it, the participles putting him at last in the full stream of it. At last (we may say) somebody in this play is alive, because Patroclus is dead: but alive on what terms? 'Crying on Hector' is what History does, as if the phrase were a synonym for *'glory'*; and the lines have something to say about human glory.

These few verse lines of near-choric commentary carry the full *Troilus* shock, and tell, as it were, the *Troilus* story. For they arouse an interest that works almost like a probe with a point at both ends, in that we find ourselves laughing at grotesquely broken creatures at their moment of greatest pain, and the laughter itself hurts. The experience is in itself a species of self-betrayal. A great deal of *Troilus and Cressida* can make one laugh hysterically—especially in performance—but never without a kind of kick-back of sadness, even of guilt: as though one were walking over someone's grave. This is a reaction to a style that must in itself have developed in consonance with the play's historical materials. For each of its two primary actions has a historical peculiarity: it seems at once distinctively distant, far removed in time, long before the stuff of the Roman plays and the English Histories, but also as near as may be to myth, to an image that speaks as directly to present consciousness as the contemporary proverb 'As false as Cressid' did. Thus, the verse most characteristic of the play makes us see and feel things both as History and as Now, as very great and very small, as far removed and detached and also as the stuff of most immediate sensation. There is direct relationship, that is to say, between the lines of Ulysses I have been discussing, and that earlier ambiguously potent moment when the lovers meet to make their affirming already disproved vow, with Pandarus beside and as it were between them, all three caught in a trap of human memory:

> When time is old and hath forgot it selfe:
> When water drops haue worne the Stones of *Troy*;
> And blinde obliuion swallow'd Cities up;
> And mightie states characterlesse are grated
> To dusty nothing; yet let memory . . .
>
> (III. ii. 188–92)

let memory do, in fact, what the play does now, 'say . . . as false as *Cressid*'. That Troilus and Cressida need stubbornly to vow a fidelity seen from the beginning as impossible proves an experience of betrayal both intimate and classic, and says something lasting about the troubling human endowment of aftersight and foresight. Both style and situation alike say that 'The Way We Live Now' (which could be the play's subtitle) is an experience as divisive as Achilles finds it, 'arming, weeping, cursing, vowing vengeance'.

'The Way We Live Now' is also a form of existence with reverberations well beyond the merely historical; conscious historicism provides a detached perspective that may itself become subject to scrutiny. If the Greeks look shallow, this is not, or not only, because the eye of memory diminishes, but because they remind us of those human beings who choose a shallow sense of History as their living criterion. 'Now play me Nestor': and Nestor does himself play Nestor, narrowing his type towards that simplicity of image which the revue-stage and the political forum alike demand. Important men, successful men, men of the moment, the Greeks—who as characters may well contain mocking reference to members of Elizabeth's court: Chapman addressed Essex as 'Achilles' in the dedication of his Homer—the Greeks embody our own consciousness that we may define importance in many different ways, a sense located in that image of the weeping Achilles and his mangled Myrmidons: giants who are children, animals, broken toys.

Troilus and Cressida often seems Swiftian, not merely in its scabrousness, but because it plays games like this with scale. Its embodying of existence lived between History and 'Now' makes us look at things, at one moment with the extreme detachment of History, at another from the sensational immersion of 'Now'. At one moment vistas grow huge and heroic—'To Tenedos they come . . .', half a dozen syllables filling a line, building a past and future—and at another everything is 'grated to dusty nothing'. The pervasiveness of the detached view in it has made the play seem as intellectual, even as philosophical as many find it; more, it has so altered perspectives as to make story-telling all but impossible (Troilus and Cressida separated almost before they have met) and it has had a radical effect too on characters and their relations with each other. It is the possibility of living as if totally composed of aftersight and foresight that makes the wholly random detached Thersites the brilliant and horrible object he is, a man whose energies are consumed by the analysis of folly in others:

> . . . and *Patroclus* is a foole positiue.
> Why am I a foole?
> Make that demand to the Creator, it suffices me thou art.
> (II. iii. 61–7)

His knowingness is operative as he first appears before us, cast as the Artist, Jamesianly 'doing' Agamemnon: 'How if he had Biles (ful) all ouer generally'. To the purely detached and external analyst Thersites, 'brain' in Agamemnon becomes only 'sense' or 'matter'—which is to say, mere pus. And it is true that human intelligence may be no more than unhealthily extraneous if unworkable and severed from life.

Matter and mind, soul and body do so drift apart in *Troilus and Cressida*, where 'Ioyes soule lyes in the dooing' as Cressida observes with bitter detachment, perhaps with more than one pun on the word *lyes*. But everyone observes in this play, and Thersites is only the extreme of the type: in the end Troilus, like a voyeur—and against all sense—patiently stands and watches his beloved drowning in her own helpless moral weakness (and Thersites watches him); and Achilles leans on a tree while the Myrmidons cut down a naked Hector.

This extreme objectivity is what gives the play its intellectual tone. But it is balanced and crossed everywhere by its opposite: an immersion in the sensational, the immediate, a function of which is the play's special interest in the *small*. This is sometimes felt to be a variety of degradation, but need not be so: the real effect is merely troublingly impassive. Feeling will for instance localize itself in a minute outward detail, when the lovers' parting kiss is for ever

> Distasting with the salt of broken teares.
> (IV. iv. 47)

Or a sudden hatred of the theatre of public and private life will create the extraordinary image of heart's blood as face-paint:

> *Helen* must needs be faire,
> When with your blood you dayly paint her thus.
> (I. ii. 95–6)

Everywhere in the play a diffused inward pain will externalize itself thus in a fierce minute perception, as if we were always being brought to consider ourselves simultaneously with a telescope and a microscope: we continually feel the intense smallness, the small intensity of the human. When

Pandarus brings Cressida to Troilus, he appraises the sensual fear that reduces her to fragility:

> it is the prettiest villaine, she fetches her breath so short as a new tane Sparrow.
>
> (III. ii. 34–5)

All the play's cool sympathy, its disgusted tenderness fuse in this image. Small birds are nice—we feel an intense anxious tenderness; but no one would compare Rosalind or Imogen to a sparrow. Nor would such heroines allow the intervention of a Pandarus whose hot-cold commentary converts human love to cruel desire.

Shakespeare has, furthermore, extended this brilliant new art of the small in a direction which differentiates this play from almost anything else he produced. He writes with an almost sociological realism, locating leading romantic characters within a specific social context—almost a class context. Both Cressida and Pandarus are—or so I think the vulgarisms of their speech indicate—what used to be called 'common'. They therefore arouse our interest, in a heroic and romantic milieu, by their unusual human specificity, they ask for our identification with them by recalling to us the fact that we too are small, are specific and localized. We are human beings that have our commonness in common. The obsession of Troilus for Cressida is heart-breaking and ridiculous, but also true, because she is not that central symbol that love looks for, but something contemporaneous, suburban, misplaced, like a character from Kingsley Amis adrift in a novel by Richardson: she has simply wandered into the wrong story. Hence the peculiar rightness—the rightness of the wrongness—of the cant phrases that help her speak what serves her for a mind, and that become the play's great off-key lines: 'You men will never tarry.' But unlikelier people catch the same idiom. The play is full of almost eerily 'modern' intonations, as when Troilus suddenly speaks with the whole modern throwaway derision of self, shrugging drily at Cressida's instruction to 'be true':

> Who I? alas it is my vice, my fault.
>
> (IV. iv. 107)

And the Queen of Beauty, Helen of Troy, makes herself forgivable for the whole Trojan war by the happy calm with which she strokes down her perfect nose and utters the immortally placid remark,

> In loue yfaith to the very tip of the nose.
>
> (III. i. 122)

But it is Pandarus who profits most from this startling new 'sociological' style. Touching and funny and brilliantly disgusting, Pandarus is the only purely *camp* creation in Shakespeare, a Kenneth Williams act somehow trapped in the Middle Ages; for he is compounded of transsexual intonations and business-like *œillades* just as Cressida resides on some borderline of class and type that one hardly knew existed in Shakespeare's time. Sexual and social ambiguities of this kind are rife in the comedies of the dramatist's period, as they are not normally—it is worth noticing—in his own; but no one else invented whole characters who embody the sensation that life itself is generically off-colour, and that if we live in society we move in 'the wrong set'.

When Troilus kisses Cressida to seal their bargain, the two of them standing within what we know for a momentary lodgement in the void, Pandarus leans forward with one of his horribly poignant cosy verbal mis-shots:

> Build there, Carpenter, the ayre is sweet.

Pandarus is rich in these modish by-words, yesterday's catch-phrases that stale as they drop from his lips; they are the poetry of a grocer-like principle of Commodity, and have their justification on the morning after the night before, as Pandarus disconsolately mutters: 'Let us cast away nothing, for we may liue to have need of such a Verse: we see it, we see it . . .' Catch-phrases are Pandarus's own attempt to 'build there', and make of him, like the 'artist' Thersites, another of the play's derisive mirror-images of the playwright. But all the play's characters similarly attempt to 'build there', to write a story other than the one we know: Achilles and Patroclus laughing in the tent, Troilus and Cressida withdrawing from the great stage, the party by night where the warriors fraternize, the long, pointless, deliberative debates that provide both Greeks and Trojans with an alternative to killing each other or dying more slowly of boredom: all are equally dignified or undignified ways of playing the losing game that for them History consists in. This sense of playing a losing game can even emerge as a kind of consciousness in the characters, conferring authority on Cressida's acceptance of her fate—'Now you haue it, take it'—and rendering classic, Pandarus's casual Sophie-Tuckerish word for mortality: 'What one thing, what another, that I shall leaue you one o' th's dayes'.

For Shakespeare too there was perhaps a losing game: the play itself. Lacking story, *Troilus and Cressida* lacks what may supply a work of literature with its defining matrix. With only the fragmented forms of narrative, the play tends always towards the amorphous, inhabiting the mind with a certain thinness; it is ill-defined, hard to see or remember in one piece. But from these losses come surprising gains. *Troilus and Cressida* is a sequence of brilliantly achieved moments that are incomparable in their power to startle, to needle, and to entertain. Without the coherence of story a play's primary value becomes its relation with reader or audience: a relation in this case surely unique in its immediacy. Self-consistent and unlike anything else Shakespeare wrote, the play can make us say to it over and over again—as Achilles said to Patroclus—'*Now* play me Nestor'.

9
'Spanish' Othello: The Making of Shakespeare's Moor

MANY studies of *Othello* confront as a vital problem what they see as some inherent randomness in the play. The current agreement, too, that the work is a 'domestic tragedy' may more tacitly voice the same reaction, depending as it seems to do on Bradley's sense of the play as less great than the others of the Big Four, because the dramatist had not fully succeeded in universalizing his materials—a judgement that brings us back to that 'randomness' again. This widespread reaction among readers and critics is not my subject here; I want to use it only to suggest that if that randomness really does survive in *Othello* as an achieved work of art, then it surely originates from the play's main source, Cinthio's prose narrative. It is hard for a reader of Shakespeare not to define literary merit as quantity of meaning—even in a case like *Othello* where the 'meaning' in a higher sense is still distinctly moot; the play, despite all the doubt, means a good deal to us. Of merit or meaning in that sense Cinthio's story has little. Given what we cannot help finding the mere externality of its avowed moral, its only meaning lies in the purposiveness of the Ensign's love-jealousy; when Shakespeare removes or blurs this he leaves what remains of the narrative as a succession of events that are 'cruel', almost in the modern sense of 'absurd'.

Thus deprived of conventional motivation the story faced the dramatist with peculiar problems. There is even evidence (as I have suggested in Chapter 2) that Shakespeare rewrote his tragedy somewhere between the stages represented by the Quarto and the Folio version, simply in an attempt to release his hero from the degradingly passive and ridiculous role imposed on him by the ruins of the original intrigue situation. This degree of essential difficulty in the story ought perhaps to make any student of the play ask what it was in the source-narrative that none the less so powerfully attracted Shakespeare as to make him decide to take it on. There may be many different answers to this question, and all of them will of course be both subjective and hypothetical: but it seems to me all the same a question that needs asking. The answer that I want to put forward tentatively here has at least the support of different kinds of evidence. I want to propose that the dramatist's imagination was compelled—and compelled at once, beyond the point of no return—by the random premiss of Cinthio's opening phrase, 'Fu già in Venezia un Moro', 'There was in Venice a Moor'. Here, surely, with the Moor who is, in the Italian, left as characterless as he is nameless but for his race-title, must Shakespeare too have begun. It is worth remarking that in the playwright's own lifetime the work seems to have been universally known not as *Othello* but as *The Moor of Venice*. And similarly, the elegy on Burbage speaks of him as the creator of 'young Hamlett' and 'kind Leer', but *not*, interestingly enough, of jealous Othello: the role is that of 'the Greued Moore', a phrase which retains both the passive stance and the race-typification of the source-story's character.[1] There may be some support too from within Shakespeare's own career for the assumption that what drew him to Cinthio's story was essentially its beginning with a 'Moor in Venice'. We do not know when the dramatist read Cinthio's narrative first, but we tend to assume he read it at least most fruitfully immediately before he began to write his own tragedy. This is a point in time from which he might look back at *The Merchant of Venice* as already an achieved success nearly a decade earlier in a busy past: when the 'Venice' it contained and the 'Moor' it presented would certainly contribute something of their own rich meaning to Cinthio's threadbare narrative, and yet were far enough behind not to hinder the emergence of an altogether new poetic possibility, in which 'Moor' and 'Venice' reacted together in their new context to make an

---

[1] The relevant excerpt is given on p. 396 of the Variorum edition of *Othello*.

original poetic world. For Othello is as different from Morocco (or, for that matter, from Aaron, further back still) as he is from the Moor of Cinthio's narrative.

The difference between the two writers is, of course, not confined to their central characters. To agree that Shakespeare used Cinthio's random story is to recognize also how much the prose story's emptiness does not hold or foreshadow the strange polarities of *Othello*, its brutal farce as well as its high tragedy, its fierce romanticism and its cool mundanity. The play is often alluded to as simple, but it is not; those critics who call it simple may differ strikingly among themselves in their very description of this 'simplicity'. Indeed, it is a leading characteristic of the work to seem simple and yet to produce very different responses from the equally sensitive and intelligent persons it has numbered among its readers.[2] An explanation of this may lie in the peculiar relation of the work to its source. What we call the 'source' of a great work of literature may never be its true source: which is instead to be found within those great accretions of experience and idea which constitute a writer's consciousness. The strength and richness of these stores of idea and experience may depend on those very qualities which tend to make them hard of access, their depth from the surface and their dislocation from each other. The value to Shakespeare of a story like Cinthio's, such as we call his 'source', may have lain essentially in its relative unlikelihood—its thinness, its simplicity, its functionalism: its capacity in short to activate by some clue or other all those otherwise unhandlable resources below consciousness; and by its lack of other merit not to obtrude on this activity once it was well begun. I am suggesting that Cinthio's 'Moor in Venice' acted as precisely such a clue, and

---

[2] The point is well made by Frank Kermode in his Introduction to the Riverside edition of the play: 'One can isolate a plot of monumental and satisfying simplicity without forgetting that the text can be made to support very different interpretations. The richness of the tragedy derives from uncancelled suggestions, from latent sub-plots operating in terms of imagery as well as character, even from hints of large philosophical and theological contexts which are not fully developed.' See also John Wain's Introduction to his Casebook selection of critical essays on the play.

that there was nothing in the impoverished narrative that follows that was talented enough to get in Shakespeare's way: the play was begun, and the dramatist stayed with it whatever inordinate difficulties occurred.

Beneath the 'simple' surface of *Othello* there are problems which are also its life, its vitality: and which it is therefore unwise to ignore. The quality that we call, in moral terms, randomness; the play's shifting from tragic mood to comic and back again; the aspect of the being and relation of Othello and Iago that has caused criticism of the play to be filled with the whole incidental debate of 'nobility' and 'ignobility'—all these underlie the play's 'simplicity'. I do not intend even to try to solve any of them. What I want to do is to suggest that they are all directly related to what I argue is the play's essential beginning: Shakespeare's acceptance of the subject of the 'Moor in Venice'. And this was so, I shall suggest, because 'Moorishness' was a condition that had a meaning, for Shakespeare and his audiences, once casually familiar though long lost to us. It was the subject of the Venetian or—more largely—*displaced* Moor which, given certain contemporary circumstances, at once fused together rich and diverse potentialities within the dramatist's mind, and called his new tragedy into being.

The true source of a poet's creativity is a subject perhaps both over-large and over-hypothetical. It can be translated into approachably smaller matters of fact by asking of Shakespeare's finished text of *Othello* a few questions so simple that it is surprising they have not been asked before. If we read the play the first word that we meet is the opening speech-prefix, *Roderigo*. Why should the dramatist have bestowed on his Venetian gull a *Spanish* name? The answer must be that Roderigo, who does not exist in Cinthio, depends wholly on his role as 'feed' (in all senses) to the character called, in Cinthio, the Ensign: here made not the friend of the Moor but his subordinate, almost his servant. The gull provides the necessary social extraversion for this underhand character now newly called Iago. Roderigo has a Spanish name, in short, because Iago has. But here a much more striking question arises. How then does Iago come to have a Spanish name?—and *such* a Spanish name, at that?

For it must be noted that Shakespeare has given his villain the same name, in Spanish, as his new King possessed: and the writer who will take pains to interrupt his tensely economical *Macbeth* with a courtly compliment to his new royal patron is not going to donate that actual or future King's name to a villain without noticing what he is doing. Furthermore, Shakespeare is careful to reiterate these Spanish names several times over in his play's first scene, which re-echoes with 'Roderigo', 'Iago', and 'the Moor', as if the writer were intent on implanting them well within the consciousness of his audience. And Shakespeare was unlikely to have been protected from the risk involving *lèse-majesté* by his own ignorance or that of audiences, for at that time 'Iago' was of all names the most recognizable both as Spanish and as James. St James was of course the patron saint of Spain, and was extensively commemorated by such shrines as that of Santiago di Compostella, in the north-west of Spain, after Rome the second largest such centre in existence: made pilgrimage to by an incessant stream of the devout, of whom the dramatist's own Helena is one, setting out as she does for 'St Jaques le Grand'. It is by considering the importance of St James or Santiago in Spain that we may light on some facts of relevance to *Othello*.

Santiago was adopted as patron saint on the basis of a handful of widely publicized though somewhat apocryphal historical events, of which the most renowned was his appearance to encourage and assist the Spaniards in the eleventh-century battle of Clavijo. The interesting thing is that this was perhaps the most significant encounter in the long struggle against the Moors; and that, so the historian of St James in Spain tells us, 'after this battle the apostle was commonly known in Spain as Santiago Matamoros, St James the Moor-killer'.[3] It seems possible, therefore, that if

---

[3] T. D. Kendrick, *Saint James in Spain* (1960), 24.

'Roderigo' came into Shakespeare's play because of Iago, then 'Iago' came into the play because of Othello—the Moor-killer along with the Moor. And, if the dramatist was content to risk the dangerous associations of the name Iago, then the reason that suggests itself is the name's affiliation to the Moor. But this is *only* so if we understand that 'Moor' could have in Shakespeare's world a peculiarly Spanish connotation. On that basis we can say that the re-echoing, in the play's first scene, of 'Roderigo . . . Iago . . . the Moor' gave to the work and its hero a Spanish resonance that nothing else could effect so briefly and successfully. Every time the name 'Iago' drops with helpless unconsciousness from the Moor's lips, Shakespeare's audience remembered what we have long forgotten: that Santiago's great role in Spain was as enemy to the invading Moor, who was figure-head there of the Muslim kingdom.

There is a limit to the amount of significance that may wisely be read into the names of romantic drama; but Juliet's 'What's in a name?' hardly stops her from dying. Shakespeare's dramatic nomenclature, in short, is interesting because it reflects certain harsh facts in the world outside the plays; and these facts help to extend that imaginative resonance possessed by mere names. During the sixteenth century Spain was the leading power of Europe, holding an eminence barely challenged by the English themselves at the Armada, and as such played a huge part in the Elizabethan consciousness; so that a 'Spanish' name would in any case sound very different in Elizabethan ears from what it would in our own; and to this general consideration we must add a very particular one. During the very years that we presume *Othello* to have been written, from 1602 to 1604, London had a ringside seat (even if an oblique one) at a crisis in the affairs of the Spanish Moors. It was in these very years that the French spy, S. Etienne, was in London attempting to persuade the English government to give assistance to the rebelling Moors of Valencia: though at length Robert Cecil was forced to decide that in view of his new King's pro-Spanish policy he could do no more for the rebels than give money and advise application to their other, because similarly Protestant, potential ally, Holland.[4] S. Etienne's attempt itself came at the end of, and is partly explained by, a sequence of years during which Protestant England, the defeater of the Armada, seems to have become something of a political asylum for refugee Moors from Spain: so much so as to produce two royal Edicts (in 1599 and 1601) effecting—for a time—the transportation of these refugees from the country, on the grounds that 'the Queen's Majesty is discontented at the great number of negars and blackamoors which are crept into the realm since the troubles between her Highness and the King of Spain, and are fostered here to the annoyance of her own people, which want the relief consumed by these people . . .'[5]

It was the common English habit from the Middle Ages on into the seventeenth century to categorize Moors as 'negars and blackamoors' (in the words of the Edict); but in fact these refugees would not have been anything that we would recognize as 'black', though a less ethnically experienced Elizabethan crowd might conceivably see olive skins and predominantly Arab features in this light. For the ancestors of these Spanish Moors, the Moors who invaded and conquered the Peninsula in the eighth century, were principally of the Berber strain: and the culture they established there, Islamic.[6] I make the point about what one must

---

[4] For the account of S. Etienne's attempt see Henry Charles Lea, *The Moriscos of Spain* (1901), 287, to which I am indebted also in the general discussion which follows concerning the position of the Moors in Spain. See also J. H. Elliott, *Imperial Spain, 1469–1716* (1963).

[5] *The Calendar of Manuscripts . . . The Marquis of Salisbury* (1906), part ix, p. 569.

[6] In the Introduction to his *Islamic and Christian Spain in the Early Middle Ages* (Princeton, 1979), 14–15, Thomas F. Glick has a useful discussion of the problem of terminology which may be applied to a very different area of study. Explaining why he prefers the phrase 'Islamic Spain' even though 'it implies . . . a contradiction in terms', to those of 'Muslem', 'Arabic', or 'Moorish' Spain, he writes: 'the population was composed mainly of Hispano-Roman converts to Islam and Berbers and there were few Arabs in the population. Moorish Spain, besides being archaic and romantic (conjuring up images from Washington Irving's *Tales of the Alhambra*) is

conceive the appearance of these Spanish Moors to be because it seems to me of great importance to Shakespeare's play. And this, not because that appearance is of any significance in itself—just as it does not really matter whether Othello is 'black' or 'tawny', an issue I shall return to in due course—but because the appearance of these Spanish Moors reflects their peculiar 'belonging' in their own country. If we visualize Othello as black, we see him as essentially standing out from the white faces around him. But the Spanish Moors who seem to have flooded Shakespeare's London did not so stand out from their countrymen. There can have been very little difference between a dark-skinned Spaniard and an olive-skinned Moor: and again, this fact is of interest as reflecting something essential about the position of the Moor in his own country. In fact, the contemporary situation of the Spanish Moor is so significant as to demand a moment's brief consideration.

In 1556 Pope Paul referred in a fit of disaffection to 'that breed of Moors and Jews, those dregs of the earth'—and the people he was referring to were the inhabitants of nominally Catholic Spain. For many centuries after the Moorish conquest in the eighth, the area we can now call Spain was a coagulation of shifting states cohabited by Christian and Islamic peoples together, with the Jews as a third and intensely influential minority. During the Middle Ages these three peoples had co-existed on terms that changed constantly, but that included an unchanging element of deep mutual interdependence, on both economic and more largely cultural grounds. There is likely to have been some degree of

also misleading on a number of grounds. Strictly speaking, Moors were the Mauri, Berbers who lived in the Roman province of Mauretania; therefore its use stresses . . . the Berber contributions to Andalusi culture. In English, Moor has racial connotations (e.g., Othello, a negroid 'Moor'; the black-moor of the standard English version of Aesop's fables) of blackness, whereas many Berbers are fair-haired and blue-eyed. In Spanish . . . the term *moro* is derogatory.'

Glick also writes (p. 3): 'Long after the enemy was vanquished, the Jews expelled, and the Inquisition disbanded, the image of the "Moor" remained as the quintessential stranger, an object to be feared.'

interbreeding during the period between the eighth and the sixteenth centuries (the Spanish royal family in Shakespeare's time was believed, perhaps rightly, to have Jewish blood); and there was also the kind of cultural fusion that makes precise understanding of what 'Spanish Moor' actually means a decidedly difficult matter, given that Jews and Visigothic Christians were absorbed into Islam at the Moorish conquest, and their descendants reconverted when Catholic conquest succeeded Islamic, and Islamic tolerance gave way to Catholic 'Reconquest'. For Spanish history from the eleventh century to the fifteenth is essentially the story of the *Reconquista*, the struggle of the Catholic kingdoms of the North to wrest the peninsula from the hands of the Infidel. And Reconquest was followed in the later fifteenth century by the imposition of Orthodoxy, as Spanish monarchs from Ferdinand and Isabella to the Philips of Shakespeare's lifetime fought to unify their great new single kingdom. Indeed, it was the very depth of the intermingling of Christian, Moor, and Jew within Spanish culture that seemed to them to dictate the new criterion of Orthodoxy: the fires of the Inquisition were lit to 'purify'.

By the beginning of the sixteenth century Spain had assumed the image it was to carry in Shakespeare's own lifetime. In 1492 Columbus discovered America; Granada, the last of the Moorish kingdoms in Spain, was finally overthrown; and all Jews who would not accept Christian baptism were expelled from the country. The three events formed one concerted nationalistic and imperialistic drive, a drive that produced Spain's 'greatness' and yet contained within itself an essential self-destructiveness. The expulsion of the Jews left the Moors the chief objects of Catholic animus. Not rich and intellectual as the Jews had been, the Moors—though their ancestors had given earlier Spanish culture so much of its brilliance—were now sunk to a mainly peasant population; but it was a numerically huge one. In Valencia, for instance, where the Moors provided most of the aristocrats' work-force, they counted for something like a third of the population. None the less, after the expulsion of the Jews the Moors inherited the fury of Orthodoxy. At first nominal baptism seemed to solve the problem, and Moors became 'Moriscos', or Christianized Moors. But by the later 1560s it was recognized that the Moors would not withdraw from their struggle to retain some vestige of their cultural identity. The last decades of the century in Spain saw bitter racial and religious strife, that gradually worsened until in 1609—a few years after Shakespeare's play was first performed—all Moors, baptized and unbaptized, were expelled from Spain.

One of Shakespeare's best-known sentences comes, as does his Moor, from Venice: 'When you prick us, do we not bleed?' That the dramatist may have imagined Shylock as a Marrano (or nominally converted Spanish Jew) one would not want necessarily to argue. But there are certain aspects of Othello's fellow-Venetian and fellow-outsider that approximate Shylock both to the Marranos and to the rebellious Spanish Moors of the dramatist's lifetime. Shylock is making, in this famous speech, a plea that he cannot himself live well by: he is arguing passionately for essential humanity in terms that allow—as Portia will show and his own extreme logic concedes—for essential inhumanity too. He is speaking, one might say, for the fierce indiscriminateness of the heart. Comparably, the tragedy of the real-life Spanish Moor was that he was, whatever his colour, in all important senses indistinguishable from his fellow-Spaniards; and this, not merely because in common practice he 'passed', he conformed to his society, but because that society was in itself infinitely unsimpler than the policy of the desperate Catholic states had to contend. Five hundred years and more of history in the Peninsula had produced a 'Spain', in the age of nationalism, that was one intense national identity-crisis, of which the Moor was essentially no more than the point of breakdown—one who like other victims would kill to defend himself, and one whose expulsion further diminished his already sharply declining country. For Spain had never really recovered from the expulsion of the Jews.

Shakespeare's tragedy opens with Iago and Roderigo, two

quasi-Spaniards by name, speaking with hatred, envy, and derision of 'the Moor'. Doing so, they call up momentarily but with intensity an element in the contemporary political situation that must have been—judging by the royal Edicts—as casually familiar to the playwright and his audiences as it is long unknown to us now. And, as Iago and Roderigo talk, it is not simply a 'black man' they are setting among 'the whites'. '*Moor*' means to Iago and Roderigo a civilized barbarian of fierce if repressed lusts; but to the dramatist himself it surely means something very different, a meaning entailed by his choice of names. The Moor is a member of a more interesting and more permanent people: the race of the displaced and dispossessed, of Time's always-vulnerable wanderers; he is one of the 'Strangers' who do not belong where once they ruled and now have no claim to the ancient 'royal siege' except the lasting dignity or indignity of their misery.

I have been trying to suggest how the story of 'the Moor' might appear if read within a world with a different mental geography from our own. In the world in which we read, America—only a century discovered in Shakespeare's time—is a great world-power, and Africa perhaps beginning to become so: to think of a Moor is to set him essentially in an African context, and to impose on him something of the history of the American coloured peoples. It is to the point that since the Romantic period Othello does seem to have been viewed within precisely this context and given precisely this history. The most valuable studies of Othello as Moor, those by Eldred Jones and G. K. Hunter, equate 'Moor' with 'African'.[7] And this equation tends to bring along with it an important subsidiary: it moves to the forefront what has become known as 'the colour question', since we think of the Moor as 'African' in his 'American' context—a black man,

---

[7] Eldred Jones, *Othello's Countrymen: The African in English Renaissance Drama* (1965), and *The Elizabethan Image of Africa* (Charlottesville, 1971); G. K. Hunter, 'Othello and Colour Prejudice', the Annual Shakespeare Lecture from the Proceedings of the British Academy (1967).

specifically, among white. I do not want to linger here in discussing the intricacies of 'the colour question' in *Othello*, beyond pointing out that Shakespeare seems to have been, in writing the play, happy to do what he does many times elsewhere, burn his candle at both ends—getting a maximal suggestiveness by implying things probably in fact self-contradictory. In *Hamlet*, the Ghost seems to come from *both* Hell *and* Purgatory; in *The Tempest*, the island seems clearly to be located *both* in the Mediterranean *and* in the mid-Atlantic. In *Othello*, the Moor is a mixture of black and tawny, of negroid and Arab; he is almost any colour one pleases, so long as it permits his easier isolation and destruction by his enemies and by himself. And here we come to what is surely the vital point: Othello's colour, which is to say his external being, is to some degree (in this work of the imagination) not a literal factor, but a matter of social assertion and reaction. He is, to repeat the phrase, 'almost any colour one pleases': and this is precisely why Desdemona, who loves him, sees his image in his mind (though in the world they live in, such inwardness of seeing may be dangerous too); and why most of the few descriptions we get of him come early in the play and are not to be trusted because they come from enemies, from the 'Spaniards' Iago and Roderigo. Roderigo's 'thicklips' (I. i. 66) is an insult aimed by a rival in love incited by Iago for sixty lines to think ill of the Moor; Iago's own 'old black ram' makes Othello's oldness and blackness only as believable as his tendency to bleat. Brabantio's first reaction is that a man who calls Othello 'a Barbary horse' is a 'profane wretch', and he himself comes to call the Moor a 'thing' with a 'sooty bosom' only when he learns that Desdemona has preferred him to her father. Indeed, it is, in my view, a particular part of the tragedy that Othello himself comes to share this hard externalism which he thinks sophisticated, and to speak of himself with a pathetic attempt at boldness as 'black'. But this is a discussion that needs space elsewhere to elaborate.

If Shakespeare himself had been asked what colour his

Moor was, I think he would have answered that few actors in his experience would permit a shade dark enough to hide the play of expression. Othello is, in short, the colour the fiction dictates. And it is in order to make this point that I have hoped to suggest that the Moor may be quite as much 'Spanish' as 'African'. It is only worth introducing some allusion to political affairs contemporary with Shakespeare in the hope of throwing light on what may have lain behind the apparent literalness of the dramatist's own allusions. The Moor is, of course, neither an African nor a Spaniard, but an actor on stage portraying the experiences of any-coloured Everyman: but our interpretation of those experiences will depend on how we read the words, and what presuppositions we bring as we begin.

I have been suggesting that Shakespeare's Moor should be seen as also 'Spanish', which is to say emerging from a situation that is as much political as ethnological—in which social relationship matters as much as colour. There is a further interest in conjecturing a Spanish background. Shakespeare adopts dramatically the situation that interested him politically—or perhaps this would be safer expressed in reverse: a writer may be attracted with peculiar sympathy towards political situations that his poetic gifts enable him to grasp and absorb. The heart of the tragedy of the real-life Spanish Moor was the ancient strength of the bonds which linked him to his fellow-Spaniards: bonds which ironically drove him (like Shylock) into a reactively defensive racism and nationalism. There is something deeply corresponding to this political situation in the way in which Shakespeare responsively *fuses* the Moorish with the Spanish, harnessing almost anything apprehended by him imaginatively as 'Spanish' to help characterize his Moor. This absorption of the 'Spanish' into his play gave it colour and substance; but more—it gave the work that puzzling multifacetedness which underlies and enriches this apparently simple tragedy. For the Elizabethan image of Spanish things itself carried with it (or so I would suggest) an inherent self-division, shadowing that crisis of identity that was the pattern of

Spanish history in the sixteenth century, at a moment which was one both of great wealth and achievement and of absolute and rapid decline. And it is, I believe, this sharply divided imagining of what it means to be 'Spanish' that helps to produce the very peculiar division of dramatic tone between tragedy and comedy in *Othello*.

To attempt to describe a whole phase of culture in a paragraph is of course ridiculous: none the less some of the most fruitful Elizabethan images appear to have been caricatures. It may merely be noted, for what suggestiveness the fact has, that in 1605, the year after that in which *Othello* was probably first performed, Cervantes published the first part of Spain's greatest single literary work. *Don Quixote* takes its power from the profound ambiguity with which it treats a certain kind of high romantic idealism, the way in which a given individual—gentle, scholarly, obsessive— treats his ordinary daily existence as a perfect Point of Honour. It does not explain the depth and richness of *Don Quixote* to say that, in doing this, it summarizes its country's inward history through the preceding century. The novel's jumping-off point is the extraordinary effect which romance in fact had on Spanish culture through the sixteenth century, serving to feed the spiritual pride of Spain with high images of the life of heroic sacrifice, the stronger for being divorced from traditional religion. When Philip II came to England to marry Queen Mary, the main pleasure of the courtiers he brought with him was to identify the sites of Arthur's imaginary adventures; and similarly when some years later the Spanish ambassador wished to describe what he saw as the villainy of Elizabeth and her government, he made his point by comparing them to characters from *Amadis de Gaulle*, the work which above all dominated the aristocratic imagination of Spain in the sixteenth century. But the very extremity and removedness of romance, and its obsession with the more external questions of Honour, made it in some way generate its opposite in Spain at the end of this period: that toughly ironic treatment of Honour in an often quite startlingly realistic urban context which characterizes the style and substance of Spain's new emerging and highly important form, the *picaresque*.[8]

Something of that romantic-picaresque polarity and contrast which must have comprised the English image of Spanish culture seems to me to have found its way into *Othello*. It is nowhere there precisely or formally localized. Nor is there any question of the Moor and Iago forming the kind of immortal twinning and pairing that we meet in Don Quixote and Sancho Panza; although it may be important that Shakespeare changed the Ensign from the friend of the hero to something parallel to the servant of the Moor. By doing so he introduces into his tragedy something of that vitally significant theme of the Master and the Man which the Spanish (in picaro stories like *Lazarillo de Tormes* and in Tirso de Molina's Don Juan play, *El burlador de Sevilla* as well as in *Don Quixote*) introduced into European literature. The horror of Shakespeare's Temptation Scene (*Othello*, III. iii) is its corruption and inversion of the Master–Servant relationship. A play too often treated as simple 'love-tragedy' is in fact impregnated with the subject of power and social hierarchies: and the Master–Servant relation of Othello and Iago compacts these meanings into Cinthio's lucid and brutal story of sex-intrigue.

These possibilities were opened up to Shakespeare, I believe, as soon as he envisaged his Moor as in some sense a Spaniard. Certain important corners of his new tragedy were at once flooded with a strange compound of the high-idealistic and the derisively picaresque. His Moor gained that wide and deep, that exquisitely painful conscience of the loss of Honour that Cinthio's Moor (by contrast) is so devoid of; Othello's imagination is enormously, preposterously vulnerable to the sense of social shame. Shakespeare's play similarly begins to find room in itself for an experience which Cinthio again knew nothing of, that derisive, ugly back-street insolence which is a reactive response to an Authority seen as at once over-absolute and unrespected. It is the

---

[8] See, for instance, A. A. Parker, *Literature and the Delinquent: The Picaresque Novel in Spain and Europe, 1599–1753* (1967).

picaresque common sense of the role (as Sancho Panza proves, in fact, a wiser governor than Quixote) that makes any reader or audience have to struggle so hard not to feel *some* sort of sympathy for that new wise underdog, the detestable Iago. And it is in part through this new 'voice from underground' that *Othello* gains its potentiality for frightful comedy, becoming at once the most romantic of Shakespeare's tragedies and the one most filled by an ugly obdurate vulgar Nasheian humour, which leaves us deeply unprotesting as Emilia, Iago's mate, calls the Moor a 'gull' and a 'dolt': for indeed Othello *is* gulled, and *does* act doltishly in this part of the play.

But there is another explanation than the spirit of Spanish picaresque for this peculiarly comic aspect of the tragedy. It has been pointed out that in creating the dramatic structure of this play Shakespeare utilized some of the forms of previous *comedy*, borrowing the scenic structures of *Much Ado About Nothing* and *The Merry Wives of Windsor*.[9] It may be similarly worth noting that two of the primary dramatic locations of *Othello*, the street and the harbour-side, are those for centuries recognizable as belonging to Roman comedy, and to the Greek New Comedy before it. In a word, an audience that found themselves at this play's opening listening to Iago and Roderigo talking derisively in an Italian street about a Moorish captain would have felt no doubt at all as to what dramatic situation they were assisting at. For Roman comedy bequeathed to Italian learned comedy (which in time passed them on to the more popular *commedia dell' arte* routines) some of the most important elements we recognize in *Othello*. Learned Italian comedy of the Renaissance was distinguished from its Latin predecessors by its fostering of a new social type and situation, that of the cuckold or *cornuto*; and it often fused this role of the deceived husband with its new translation of a (dramatically) much older type, one found not only in Roman comedy but in the Greek before it—that of the braggart soldier. What makes Othello's 'Spanishness' of striking relevance here is that in the world of

[9] Emrys Jones, *Scenic Form in Shakespeare* (Oxford, 1971), 121-7.

Italian learned comedy (and in popular comedy after it) this braggart who is often the deceived husband is also most characteristically a new national type: the *Spanish* soldier of fortune. For, as Boughner records in his valuable study of this character type in Renaissance comedy, 'Latin drama . . . was precisely the vehicle needed by the Italians for their mockery of the pitiless Spanish mercenaries that swept over the Peninsula in the sixteenth century and shook its civilisation . . .'[10] The braggart soldier in this guise became a directed Italian protest against the invading Spaniard, the 'barbaris hostis Italiae', 'tam ineruditus quam inflatus superbia gothica'; and he was re-imagined for these comedies in a quite new guise as a pedantic and fantastic grandee of Castile, who added to a gravity of demeanour and decorum of speech and gesture, a peculiar elegance that was believed to derive —as did so many civilized Spanish things—from those 'womanish men', the Moors.

The sense in which Othello is *not* a Spanish braggart captain will be obvious to any sane and sensitive reader of the play. In this there is an obvious contrast between him and one of his other sources or prototypes, Morocco in *The Merchant of Venice*, whose boasting oath 'by this scimitar', and threat to 'outstare' and 'outbrave' set him well within the comic braggart type, and help to balance Portia's tartly racialist revulsion from him. And yet it would not have been surprising if some of the play's first audience, finding themselves listening to a soldier and his gentlemanly gull—a gull who might be straight out of a city comedy—both of them with Spanish names, and talking, in these back streets of Venice, of an apparently supremely arrogant Moor, were not a little disappointed to find the Moor so *little* a braggart; and did not mutter, like Rymer later, that the play was 'a bloody farce, without salt or savour'. For only a certain grimness, a lack of the lightweight in Iago's intense tone, differentiates the circumstances at the play's beginning from those of scores of Italian learned comedies of the Renaissance.

[10] Daniel C. Boughner, *The Braggart in Renaissance Comedy* (Minneapolis, 1954), 20; and *passim*.

We might be in at the start of just such a comic-romantic story of jealous love as Bentivoglio's *Il Geloso* or Gabiani's *I Gelosi*, two among the many plays which such experts on the subject as Boughner or Marvin Herrick (in his *Italian Comedy of the Renaissance*) class as absolutely typical and trivial representatives of the Spanish-braggart plays of the period. And, far enough away as these two comedies are from the enormous depth and power and meaning of Shakespeare's tragedy, it is a fact that *Othello* contains devices that seem a distant disturbing ironical echo of braggart conventions which two such trivial comedies exemplify. Behind, for instance, Othello's own wonderfully romantic and just possibly ironic rehearsal of the story that won Desdemona, the enigmatically splendid account of his heroic travels and battles, there lies the braggart's invariable evocation of the grandeur of his travels and campaigns: as Zeladelpho in *I Gelosi* boasts in his prose declamation of prizes won by scattering enemies protected by hundreds of cannon, of illustrious friends and patrons, and of campaigns and travels in faraway Africa, Egypt, and Mesopotamia; or as the braggart captain in *Il Geloso* has his verse peroration concerning his achievements in Tunisia, in Barbary, in Vienna and Hungary, interrupted by the jeeringly undercutting echoes to his boasts by his valet, Trinchetto.[11]

[11] Vincenzo Gabiani, *I Gelosi* in *Commedie diversi* (Ferrara, 1560), 28^r: 'Tu dici il vero, che i priegiati, & horrevoli arnesi sogliono far riguardevoli i Capitani. Ma che mi curo di quello io havendo gia acquistato il credito, & fatto la riputatione? per havere condotto a fine tante imprese, & maraviglie, come fa il mondo. Senza che gli arnesi non sono quelli, che mettono i pari nostri avanti, appresso alle corone, & a gli scettri. Ma questa quà si bene, che importa il tutto. Va domanda in Acarnania, in Egitto, in Soria. Domanda di me in Aphrica, in Guascogna, in Boemia, & sopra tutto i Mesopotamia, et sentirai la relatione, che te ne sarà fatta.'
Hercole Bentivoglio, *Il Geloso* (Ferrara, 1547), 18^v:

> O quante
> Altre gran prove hò fatte ch'or non dico,
> Che non è tempo: a Tunisi che feci
> Di Barberia? che feci ancho a Vienna,
> In Ungheria? non presi non uccisi
> Un numero infinito di quei Turchi
> Con questa spada . . .

Such echoes may be fortuitous. Marvin Herrick's wide-ranging study makes many links between a number of Shakespeare's comedies, and some of his tragedies, and both the Italian learned comedy and its popular successor, the *commedia dell' arte*, but finds *Othello* one of the few plays by Shakespeare not worth considering in this context: he simply fails to mention it. And yet it seems to me a detail striking enough to need some consideration that one of these two trivial comedies, Bentivoglio's *Il Geloso*, provided Ben Jonson with the characters who were the ancestors of his Bobadill and Kitely, and that one critic has suggested that it was from this very play, Jonson's *Every Man in His Humour*, that Shakespeare may have found the basis, in Thorello, for the name he gave his Moor, Othello.[12] The link at any rate adds to the materials for believing that these Italian learned comedies, in which the figure of the Spanish braggart was a principal attraction, were an important feature of that half-tragic and half-comic world that sprang to life within Shakespeare's energizing and unifying imagination. Already his Don Armado in *Love's Labour's Lost* had shown how far an innately rich and delicate sensibility could refine the merely dramaturgical device of the coarse braggart into something at once far more truly 'Spanish' and far more individually Shakespearian. For Don Armado has something of that helpless imaginative refinement, that rigid vulnerability to idealism, which ten years later was to make the Don of Cervantes the great—the of course much greater—classic he remains. (And it may have been in response to the divided vision of Spain that Shakespeare impassively gives to Don Armado a servant-girl for a Doña, as Cervantes was to do with *his* knight).

Othello also and much more darkly seems to reflect this double sense of what it might be to be Spanish: an experience of tension between a fastidious romanticism and an earthy and sometimes brutal directness. Certainly there seem to me to be problem areas in the play which cease to be problems

---

[12] Emrys Jones, op. cit. 149.

when seen simply as one aspect or another of this divided experience. So one might consider, for instance (and I mention here only a random handful of cases, differing in interest and scale), the strangely wordy gauche refinement, straight out of Don Armado, with which Othello himself anxiously denies on the day of his elopement that he could ever be subject to desire or 'heat, the young affects | In my defunct and proper satisfaction' (I. iii. 261—74, a speech that needs discussion though it has never to my knowledge had it: Othello's embarrassment actually creates verbal crux); Cassio's inexplicably intense and silly romanticism (II. i. 62–82), emerging from a character for whom Shakespeare has invented a whore for him to keep company with; the calm, social acceptance with which Desdemona follows the practice of earlier *comic* heroines in chatting with a clown, joining as she does in Iago's unfunny badinage at the harbour-side; and most of all, the unerringness with which for dramatic reasons we find ourselves at once agreeing to complicity with our detached, comic guide Iago, who on all human grounds is boring, shallow, vicious, and in no way whatever to be trusted.

All these are aspects of the play which seem wholly right in their context, and yet which continue to puzzle if we impose upon the tragedy some over-simplifying category. All are facets of the one central premiss, and are necessitated by that originating idea which fused together in Shakespeare's imagination great diversities linked only by the code-word 'Spain'. Seen from any other angle, Cinthio's story offered Shakespeare scarcely anything but that meaningless line of intrigue-narrative which the tragedy holds on to with an impassivity in itself contributive; everything else, including the meaning, Shakespeare found for himself. But it was the intrinsic 'Spanishness' of that Moor-in-Venice opening ('Fu già in Venezia un Moro') that had begun his second great tragedy for him.

I hope that I do not seem in the foregoing to have argued that *Othello* is (as Rymer suggested) a comedy; or that its characters are in reality of Spanish birth, or that its hero is a braggart, or that he is black (or white). The intention of this essay has been merely to ask some questions about the formative period of one of Shakespeare's most brilliant plays: that phase of reflective reading-around while the dramatist was beginning to invent a new work. In doing so, I have had both a negative and a positive purpose. Negatively, I hope to challenge our perhaps too simple 'African' sense of Othello. For a century and more we have tended to see Shakespeare's play in the light of certain deep even if tacit or indeed unconscious post-Romantic presuppositions which in fact derive from a more or less modern myth of the Moor—the Moor as essentially 'African' or 'black', in both a literal and a metaphorical sense. We have thus come to see Shakespeare's play, or so it seems to me, as almost indistinguishable from a work that shares these (as we may loosely call them) Victorian presuppositions: we see it as much like Verdi's opera *Otello*, as a work that is simple, beautiful, full of passion and of pain, lyrical and barbaric and above all, all about Love.

Shakespeare's play does have some connections with this image: but the image is far from a wholly true one, and as such may silently distort and confuse. It is in an attempt to supplement that too partial image of the play as about an 'African' Moor that I have tried to suggest that Othello is in fact 'Spanish' as well. And this is a matter which reaches back beyond the purely political context of Shakespeare's own time into a great literary background that is vital to the play. If one deprecates Victorian romance in the consideration of the play, this is not because it is bad in itself but because it may serve to conceal that great world of Renaissance romance which is not precisely the same thing, but which surely contains some of the true sources of *Othello*. It is a curious fact that the sole proof—if it is proof—that Shakespeare read at least some Ariosto in the original is located in Othello's phrase about the Sibyl's 'prophetic fury'. It would not be surprising if the *Orlando Furioso*, that great source of the Moorish for Italian learned (and hence popular) comedy, taught much to a writer far greater than those comedies could provide. The background to *Orlando Furioso* is the perpetual, dream-like war of Christian and Pagan, and the Pagans are Moors, the wars being waged by the Kings 'of Affrike and of Spayne'—this last a phrase that re-echoes memorably through the poem; Rogero, for instance, the inamorato of one of the two heroines, and a heroic Pagan whose colour is immaterial but clearly not black, is referred to as a 'knight of Affrike and of Spaine'.[13] His final conversion and marriage to Bradamante concludes reasonably enough this great chronicle of romantic courtesy that begins with the famous tender, slightly ironical, image of the two 'auncient knightes of true and noble heart', one Christian and one Pagan, sharing one horse 'like frends'. Without stopping to consider whether or not the poem might be called another of the play's sources, one can say at any rate that this is surely the world that Shakespeare's Moor—who is 'of Affrike *and* of Spaine'—in some sense comes from, and in another sense would dearly like still to belong to. But Othello is *not* a knight, but a mercenary; and the realm he serves is not Ariosto's dreamily Charlemainean landscape of the past, but 'present-day' Venice, the great trade city—where, as the opening lines of the play make grimly clear, to be a 'frend' is to have 'my purse, | As if the strings were thine'. It is thus that we may say again that when Shakespeare read in Cinthio of a '*Moor in Venice*', his tragedy was begun.

[13] *Orlando Furioso*, tr. Harington, ed. Robert McNulty (Oxford, 1972), Book 1, st. 6; Book 30, st. 70.

10

Two Damned Cruces: *Othello* and *Twelfth Night*

THE establishment of Shakespeare's text is now a matter much conditioned by technology. Yet it remains deeply involved with literary criticism. Any given crux will demand an assessment of what surrounds it; and the attempt to solve even the smallest of textual problems can involve and then enlarge a reader's understanding of the entire literary work. In the first twenty lines of *Othello* we meet a case of this—a word, and indeed a whole line, that constitute one of the best-known and longest-unresolved cruces in the canon. The play's two original texts, the Quarto and the Folio, give a version of the passage that differs in one or two significant details, though the problem stays much the same in each. Iago, apparently lucid but showing signs also of an incoherent rage, is attempting to pacify an angrily jealous Roderigo by asserting his own hatred of the man who has won the place he himself wanted as the Moor's Lieutenant:

> and what was he?
> Forsooth, a great Arithmetition,
> One *Michael Cassio*, a Florentine,
> A fellow almost dambd in a faire wife,
> That never set a squadron in the field,
> Nor the devision of a Battell knowes,
> More than a Spinster . . .
>
> (Q 20–6)

So the Quarto, with *s* and *v* modernized (as I have done with both original texts, here and throughout what follows). The Folio reads:

> And what was he?
> For-sooth, a great Arithmatician,
> One *Michaell Cassio*, a *Florentine*,

> (A Fellow almost damn'd in a faire Wife)
> That never set a Squadron in the Field,
> Nor the devision of a Battaile knowes
> More then a Spinster.
>
> (F I. i. 20–6)

Johnson referred to the fourth line here, describing Cassio as 'damned in a fair wife', as 'one of the passages which must for the present be resigned to corruption and obscurity'. His 'present' has lasted until the present day. Editors continue doubtfully to reprint the line, which still reads as nonsense. And most of them would probably agree with the invaluable if now hundred-year-old New Variorum edition, which gives to the line five pages of notes (in very small print) ending with the echo of Johnson's '"I have nothing that I can, with any approach to confidence, propose"—Ed.'

Earlier editors occasionally emended either 'damned' or 'wife', without much improvement; but most have tried to make sense of the line not by changing it but by glossing it imaginatively. Many commentators guess at some altered intention on Shakespeare's part to marry off his Cassio, whose representative in the source-story was himself married; or they suppose here some allusion to Cassio's future affair with the whore, Bianca. It is in general believed that the line designates the Lieutenant a ladies' man.

None of these proposals meets the real problem of the line: that it fails to make sense. Productions of the play which drop the line do in their way comment on it more effectively than most editions. And the failure to make sense is intrinsic. Even if we should (and it seems a mistake that we do) call in a wholly hypothetical or unconjugal as yet unencountered mistress, it remains a fact that no one was ever *damned* for having a fair wife; and to be 'almost damned' is almost as curious a concept. Some sense of this unusual, not to say Calvinistical approach to damnation must account for the regularity with which editors cite an Italian proverb of the time, 'L'hai tolta bella? Tuo danno'. This, M. R. Ridley, in his otherwise excellent New Arden edition, firmly translates,

'You have married a fair wife? You are damned.' His sixteenth-century Italian is, it must be said, very likely to be better than the present writer's. But the proverb certainly looks as if it means 'Your girl's good-looking? That's your loss'—with no mention of damnation; and Florio as certainly translates *danno*, in his 1598 *Worlde of Wordes* (an Italian-English dictionary), '*hurt, losse, danger, dammage, perill, skath*'—again, with no mention of damnation. The two closest English equivalents of the time given by Tilley's dictionary of Renaissance proverbs are: 'He that has a white horse and a fair wife never wants trouble', and 'Who has a fair wife needs more than two eyes': yet again, no talk of damnation here.

The problem of meaning here presented by 'damned' is reflected linguistically in a problem of grammar. The phrase 'damned *in*' does not appear to exist—in so far, of course, as it means 'damned as a result of something' (one may speak of the damned in hell). In the several columns given by the *OED* to 'damn' and its derivatives, 'damned', 'damnable', and 'damnation', nowhere can be found examples of these followed by *in*. All have some tendency to be used absolutely and without prepositional additions, but where these words do govern a preposition it will be *for*. 'Damned in' does not occur. We accept the phrase unthinkingly as a Shakespearian nonce-usage perhaps because the speaker, Iago, has such a compelling and question-begging presence; partly too perhaps because, if we know the play already, we may hear the line as somehow relevant to the Moor himself; but most of all because the word 'damned' has in itself a hypnotic power, whether through its theological absolutism or merely its expletive energy. This silencing force may surely be presumed to have worked on the play's first printers. That the same mistake—if a mistake there is—should feature in both of *Othello*'s two early texts is not in itself surprising. If the word underlying 'damned' was obscure or illegible enough to fox one compositor it was difficult enough to bewilder the other. Nor is it strange that both the Quarto and Folio texts should agree in reading 'damned'. If we assume that the Folio text was printed from a Shakespearian manuscript which was a twin to the source-manuscript for the Quarto *Othello*, but which had since the first production of the play received substantial Shakespearian revisions and additions, then the second or Folio compositor might well, if uncertain of his manuscript reading at this line, turn to the printed Quarto text which he used for a guide. The result would be 'damned' in both texts, however differently spelled.

It seems clear that 'damned' has, like all such words of power, exerted a kind of hypnotic force on all the play's editors, as surely as on its first compositors. For the notion of someone being in any sense (whether expletive or theological) *damned* retains the power to attract and *dis*tract; obstructive and interruptive, the word stops readers from noticing that Iago's discourse here has a readily comprehensible flow. But there is something else again that hinders the easy reading of these lines. Modern editors usually retain at least some of the thicket of brackets that feature in the Folio text of *Othello*, on which most current editions have been based—that feature, that is, at least in some (mainly early) scenes, and in Iago's speeches in particular. Hence the brackets that in the second version here enclose the 'faire Wife' line, decisively isolating it from the rhetorical and syntactic movement of the whole. Although they seem not to have attracted comment from textual critics, these curious, irregularly clustering brackets are interesting, and one would not want simply to dismiss them. They can strike a reader of the Folio text of the play as suggestion, however faint and fanciful, that the dramatist may have come to see his villain as possessing a kind of wholly fake 'Jamesian' quality. Iago, in fact an entirely social and superficial personage, conceals his emptiness from himself and others by an external show of inwardness, of secret, reflective depths; he talks endlessly in parentheses and hypotheses. Brackets are, as it were, the spirit of the man—Iago's soul. This is a concept which the actor of Iago may have always found himself intuitively noting, meeting the character's natural speech-rhythms with grunting sinkings of the voice, a habit of throwaway confidentiality. But it is a speech-rhythm anyway liable to thin out, like all such mannered indices of characterization, once the play is well under way. And, from the reading point of view, it is not easily practicable in a text with Elizabethan punctuation systems. Both playwright and printer alike may have sometimes been uncertain exactly where the parentheses, in such an in fact highly public rhetoric, ought to close themselves. In this sense brackets round the 'faire Wife' line may falsify, and the Quarto reading prove more reliable in that it gives the line back to the context to which it belongs.

That context is most clearly seen by removing the dangerous, because mesmeric, word 'damned', together with the preposition 'in' which it governs, and then rendering the rest into modern English without punctuation. Iago answers his own rhetorical question, 'what was he?':

> One Michael Cassio a Florentine
> A fellow almost   a fair wife
> That never set a squadron in the field
> Nor the division of a battle knows
> More than a spinster

Set out thus, it becomes evident that, despite some syntactical 'clotting' produced by the incoherence of the speaker's rage, the 'fair wife' is an operative phrase and governs 'That never set a squadron in the field', just as the 'spinster' 'knows [not] the division of a battle'. And the 'fellow' is compared to both, just as the 'fellow' itself is in apposition to 'Michael Cassio'. This is only hard to apprehend in so far as Iago wishes it to be, introducing a confusion parallel to that 'almost' in its capacity to shield the speaker from the blame of what he is saying. As Iago uses his syntax as a shield, so does his vocabulary here play tricks that need watching. One of the reasons, I believe, why this 'crux' has stayed a crux for centuries is that Shakespeare has given to Iago language with a self-defeating element: his terms are so much involved in the social medium of the period as to be historically vulnerable—they are half-lost to us now.

Iago is in fact creating the impression of plain statement

(which is why editors struggle to give Cassio some kind of 'real' wife) while actually fabricating a deceptive texture of insults. This sleight-of-hand starts with the word 'Florentine'. Why does Shakespeare choose to make his villain mention Cassio's place of origin? The answer, I think, lies in what Iago does with that factuality: quietly converting simple datum into implication, and thence into insult. Because Roderigo is violently disappointed and he himself viciously angry, Iago's citation of a mere place of origin makes it work to give Cassio a taint of the outsider, the 'stranger' to the two Venetians on-stage: a foreshadowing, of course, of what will be done to the Moor. In its context 'Florentine' manages to sound, retrospectively, like a curiously dirty word: a shift induced by Iago's continuing his alliterative *f*s. But 'fellow' and 'fair wife' would communicate something other and more to an Elizabethan ear than to ours. We hear in 'fellow' only some amiable, jocose, and Edwardianly dated expression of equality of status with another male; and 'wife' holds for us simply the sense of conjugal relationship. But in Shakespeare's day, and after it right up into our own immediate past, 'fellow' was unmistakably a socially conditioned word, used by the nobility or gentry only to or of a person socially inferior. As to 'wife', the marital relationship is the second meaning offered by the OED; the first and earlier usage meant 'woman' only. Phrasally ('a wife that did costerds sell', 'the wyfis that fostred yow') this came to be applied to tradeswomen of low social rank (a use retained now only in such compounds as 'fishwife'). Given its relationship to such phrasal uses, and its alliterative context here, Iago's 'fair wife' carries a hint of something almost approaching 'nice little piece'.

I am suggesting that the 'fair wife' does not exist as an element in Cassio's personal history, but serves as a descriptive definition of some flaw (as Iago sees it) in Cassio's character. As the 'spinster' locates the Lieutenant's ignorance of 'the division of a battle', so the 'fair wife' establishes inability to 'set a squadron in the field'. Moreover, 'spinster' is a term as potentially derogatory as 'fair wife', and on the same grounds. Editors still sometimes insist that in Shakespeare's time 'spinster' had no necessary female connotation, but simply indicated a (male or female) spinner by trade or occupation; yet all the Elizabethan examples cited by the *OED* either plainly indicate females or leave the gender unstated. Consonantly, the unbiased ear of a reader surely hears in the present passage a kind of rhyming, or vituperative chiastic patterning, of the triumphant closing rancour of 'spinster' as against the opening 'fair wife'. If there is some shadow of contempt in that 'fair wife', then so too is there in the spitting conclusive 'spinster', which carries a charge, as of 'withered old virgin', as firmly as the earlier 'Florentine' comes to sound absurd, even indecent. In short, the use here of 'spinster' may provide a small instance of Shakespeare's linguistic brilliance. He may have so caught a socially evolving word as to make his Iago the first person in literature to turn the estate of female celibacy into an insult in itself, and at the same time to make the comparison of a man with a woman an insult in itself, and to do both so ambiguously as to be all but undiscoverable.

For Iago's purpose is dangerous enough to need the covering protection of that 'almost', and all the opacity of his furtively vivid rhetoric. If the 'fellow' is being compared equally to the 'fair wife' and to the 'spinster', then the obvious point of their comparison is their shared femaleness. The major intention of these few lines is to revenge Iago on Cassio by diminishing him: specifically, by the imputation that male intellectuals of his kind ('Arithmatician' is made to sound another curiously dirty word) are of course effeminate. It seems to me that this suggestion is maintained through the allusion that immediately follows these lines, an immediacy accentuated in the Quarto (which I here quote) by the more open and fluid punctuation:

> Nor the devision of a Battell knowes,
> More than a Spinster, unlesse the bookish Theorique,
> Wherein the toged Consuls can propose
> As masterly as he . . .
>
> (Q 25-8)

TWO DAMNED CRUCES

The word 'toged' is counted among the play's many cruces, for the Folio here reads 'tongued'. It may be that the dramatist decided to drop 'toged' (togèd or toga-ed) because the deviousness of its allusion proved difficult for both actor and audience to cope with; but that he had hoped to let Iago carry through it his snide hint of an intellectuality seen as absurdly womanish as well as 'bookish' (consuls wear *togas* instead of decent *macho* breeches) in contrast to the male and properly experienced soldiers. Certainly Iago returns to this theme later (II. ii) and there more broadly and coarsely—it is the more openly rammed home. Pretending to defend Cassio as far as he can for his attack on Montano, Iago speaks with a plain man's brutal kindness which (in fact) manages to hint for Cassio an even darker, more diminishingly indecent situation than the one he has on his hands as it is. The Lieutenant is presented as if caught in some sudden violent quarrel of randy rough-trading homosexual lovers (an imputation that simultaneously casts a shadow over the other 'Bride, and Groome', Desdemona and Othello):

> Friends all, but now, even now.
> In Quarter, and in termes like Bride, and Groome
> Devesting them for Bed: and then, but now:
> (As if some Planet had unwitted men)
> Swords out, and tilting one at others breastes . . .

(I quote here from the New Variorum Folio text: F II. iii. 203–7).

The general tendency, then, of Iago's rhetoric in this opening passage of *Othello* makes itself clear. He is intent on darkening and diminishing Cassio, vengefully, in any way he can. The way that offers itself most immediately is the 'little woman' implication—a hint which is the uglier obverse of that image of Cassio that troubles Iago by having a 'daily beauty in his life'. This intention is reflected in the actual phrasing of the passage. The issue is the comparison of Cassio and the 'fair wife'. Therefore, in a line that reads 'A fellow almost   a fair wife', the missing words that underlie 'damned in' must mean something like 'like', 'equivalent to', 'describable as', or 'portrayed in'—indeed the concept of the *portrait* is desirable in that Iago's 'almost' suggests an image so alarmingly precise and definite as to need some escape-route into indeterminacy. The word or phrase, moreover, needs to be of that degree of difficulty that could explain its misreading by a printer as expert as Shakespeare's at this stage can be expected to be.

This provides enough information to make a guess. But further information, making guessing almost unnecessary, comes from a different source. I want here deliberately to pause, and go on a loop-line of argument that will take us back to the play written only two or three years before *Othello, Twelfth Night*: for it is there that we find a second crux so involving the word 'damned' as to suggest, by its solution, a possible form for the word in *Othello*. Two textual cruces so curiously linked should at least be allowed to throw what light they can on each other.

The third scene of the first Act of *Twelfth Night* introduces the two knights to us. Much of the scene is taken up by Sir Toby's mockery of Sir Andrew's keen but incompetent wish to engage in the great world of life and love, a wish encapsulated in his innocent passion for the dance. This rallying is brought to a head (or perhaps a foot) by Sir Toby's amused flattery of Sir Andrew's 'legge', a compliment which the other accepts with gentlemanly detachment:

I, 'tis strong, and it does indifferent well in a dam'd colour'd stocke. Shall we sit [set] about some Revels? (I. iii. 126–7)

I quote here the New Variorum edition of the play's single original text, the First Folio. This 'dam'd' of the Folio, though a much-discussed crux, does not excite quite such editorial debate as the 'dambd/damn'd' of *Othello*. In early editions it was sometimes allowed to stand, on the assumption that 'dam'd colour'd' means 'hellfire coloured', which is to say 'flame red' or (a minority verdict) 'black'. But most commentators have felt this to be a strained interpretation, and surely rightly; to mean this the text would have to have read something more like 'damnation coloured', and even then it is uncertain that a Renaissance mind could define a colour in this way (there is an instructive contrast in the relative concreteness of *Love's Labour's Lost*'s 'black is the badge of hell'). Most texts therefore emend 'dam'd', with proposals including 'flame', 'damask', 'dove', 'paned', 'claret', and 'dun'. Of these, Rowe's 'flame-coloured' was for long the most popular reading, although more recent editions appear to be replacing it with Collier's 'dun-coloured' (and the change may be in itself an interesting reflection of our whole altered sense of the play. The consciously darker and subtler shade has replaced the brighter and prettier image.)

Yet all these emendations lack the sense of necessity. It is charming enough to visualize Sir Andrew in either a bright red or a dull brown stocking; but both alternatives have that pure post-Romantic randomness that is perhaps not a part of the character of Renaissance literary art. Detail in Shakespeare is invariably pointed (Shylock's ring, Lear's button). The impression that a point is being made here, which neither 'flame' nor 'dun' meets, is underlined by a small fact oddly ignored by editors. Sir Andrew can hardly be merely referring to what he wears at that moment on-stage, which would be artlessly tautologous in a way that is not explained away by his folly; he must be indicating some altogether other pair of hose, one reserved for and even symbolic of high days and holidays—an essentially 'special' stocking.

Stockings recur in *Twelfth Night*—they could almost be called a motif or symbol in it. It therefore seems unlikely that Sir Andrew's allusion to an off-stage wardrobe has nothing at all to do with one of the play's most amusing events, Malvolio's assumption of yellow stockings in the name of love. One of the rather few things we appear to know about Elizabethan hosiery is what seems to have been a vogue for yellow stockings at one period, and their survival for a longer one as a kind of (often ironic) talking-point. The New Variorum prints a note by the Victorian critic W. A. Wright, glossing Malvolio's quotation from his pseudo-love-letter from Olivia, 'Remember who commended thy yellow stockings' (II. v. 144)—and the whole note is informative enough to be worth quoting here:

'Yellow stockings' were apparently a common article of dress in the 16th century, and the tradition of wearing them survives in the costume of the boys at Christ's Hospital. They had apparently gone out of fashion in Sir Thomas Overbury's time, for in his *Characters* [1614] he says of 'A Country Gentleman', 'If he goes to Court, it is in yellow stockings'; as if this were a sign of rusticity. They appear to have been especially worn by the young, if any importance is to be attached to the burden of a song set to the tune of *Peg a Ramsay* ... in which a married man laments the freedom of his bachelor days: 'Give me my yellow hose again, Give me my yellow hose'. Malvolio may have affected youthful fashions in dress.[1]

From all this and other such references we may gather that yellow hose became, either in fact or as a literary theme of the time, a kind of focusing point between three areas: fashion, the Court, and love—and that like other fashions and ideas about fashion, they became too an index of staleness, the fabric of a bad joke. A tune like Noel Coward's 'Somewhere I'll find you' (instead of *Peg a Ramsay*) might still evoke both a vivid ghost of romanticism and an allusion to the 'potency of *cheap* music'—the charm of the modish and the repulsion of that charm. In the same way, Malvolio's investment in yellow stockings may bring with it all the by now (in 1601 or 2) accreted ambiguity of the fashionable courtly loveromanticism, a mode always 'out of date', distasteful, to those of true refinement and sensibility. The stockings act as a near-symbol of love à la mode, a passion ridiculously attractive to worldly climbers like Malvolio, but absurd or disgusting to the sincerely feeling like Olivia ('a Cipresse, not a bosome, | Hides my heart').

Seen thus, Malvolio's yellow stockings serve to highlight an essential element of *Twelfth Night*—one perhaps too little stressed by criticism, though brought to the fore in a brilliant production of the play a few years back by Jonathan Miller. This is its essential courtliness. The comedy holds at its heart

[1] The original note is on p. 128 of William Aldis Wright's edition of *Twelfth Night*, Clarendon Press Series (Oxford, 1885).

the question of the value and meaning of the high-Petrarchan romantic love most at home in a Court. This courtly context lies behind even the fooling of the two knights in the third scene of the play. Sir Andrew, a character often gulled into farce, at the same time derives most humour and most interest from the natural gravity (surely) of his expression; a formal man, he is like a young and highly proper Foreign Office underling, decorously agog to do the right thing—which for him means the smart thing, even the courtly one. This desire here extends even to the anxious propriety of his garments. He announces nonchalantly his ownership of the right colour stockings in a manner that even imitates the Court's *sprezzatura* casualness: 'It does indifferent well . . . '. In the process he tells us something about both the comedy and himself, and establishes a context for the gulling of Malvolio later on.

But the ambitious Malvolio is a vulgarian, and Sir Andrew is—though weaker and much more stupid than the Steward—pathetically 'well-born', vulnerably a gentleman. It seems possible that Shakespeare saw Sir Andrew as wistful and silly but refined enough to opt for something a little above 'yellow' in stockings. At a period in which one can come across a reference to 'peach-coloured' stockings, Sir Andrew may have favoured the special, sharply luscious shade of yellow that one calls *lemon*.

To see Sir Andrew's hose as 'lemon-coloured' rather than 'dam'd colour'd' has more justification on textual grounds—as well as literary—than any of its alternatives. Florio's Italian–English dictionary, quoted earlier, translates the Italian *lemone* as 'the fruite we call a lymond', and *lemonino* as 'a kinde of lymond colour'. It seems worth conjecturing that Shakespeare wrote his phrase in a form very close to Florio, as 'limond colour'd'; and since neither the word nor this spelling would be particularly familiar to the compositor, he would guess at the substitute to which, in fact, 'limond' is closer than any of the proposed alternatives, the word 'damn'd/dambd/dam'd'. For in secretary hand the introductory stroke of the *l* of *limond* could resemble the bowl of a minuscule *d*. On the other hand, the word 'Lemon' occurs later in *Twelfth Night*, at II. iii. 28, and is in its original text, the Folio, printed with a capital *L*. And the relevant fragment of *Sir Thomas More* (which Shakespeare may or may not have written) shows that a majuscule *L* followed closely by an *i* could even more easily be read as majuscule *D*.[2] In addition, many textual cruces of the time arise from the fact that this style of handwriting reduces any series of *i*, *m*, and *n*, and sometimes—in their neighbourhood—*o* and *a* too, into a sequence of indeterminately numbered, small, wave-like strokes.[3]

'Lemon-coloured' or 'limond colour'd' gains support from a rather different kind of evidence. I have just mentioned that in the second Act of *Twelfth Night*, Sir Andrew enquires of Feste: 'I sent thee six pence for thy Lemon, hadst it?' Editors are agreed that the knight means not the fruit but the girlfriend of Feste, here referred to grandly as a 'Leman'. If Shakespeare, in his reference to the stockings, though spelling the word 'limond' still pronounced it 'lemon' and could hear in it the 'lemon/Leman' play implied by the printer's error later, then Sir Andrew's meditation on his hose is thereby enriched. For it happens that this allusion to 'lemon-coloured stockings' is immediately followed (as the occasional Victorian editor disapprovingly noted) by some bawdy verbal by-play on the part of Sir Toby. Though the comedy as a whole has little innuendo of this kind, perhaps because of its Court-context and the high-serious treatment of love, what there is tends to cluster in this present scene, I. iii: where Sir Toby and Maria are in effect trying to initiate Sir Andrew into a more manly role in the great dance of life and love—an attempt that comes to a climax in this otherwise pointless piece of backchat that takes the two knights off the stage. Sir Toby, insisting on the part played in

---

[2] See Harleian MS 7368 fol. 8(b) line 11 for majuscule *L* as the initial of a verb.

[3] I should like to thank Mr R. E. Alton for his valuable palaeographic suggestions. I am also indebted to Dr Stanley Wells for some relevant textual considerations elsewhere in this article.

our conception and birth by 'legges and thighes', adds 'let me see thee caper. Ha, higher: ha, ha, excellent'.

The undermeanings of these exchanges may be understood as depending (as of course 'legges and thighes' do) on the suggested allusion to 'lemon-coloured' stockings. Because of the 'limond/lemon/Leman' fusion, Sir Andrew articulates a muffled impression that 'Leman-coloured' stockings are not only smart but positively sexy. Since the whole modern fashion industry profits by something like the same illusion, he can hardly be blamed for it.

I left *Othello* on the suggestion that we need to find a substitute for 'damn'd': a word which the context requires to mean something like 'describable as', 'portrayed in'. It may be that *Twelfth Night*, close to the tragedy in its time of composition, will offer information about the form of the word we are looking for. If Shakespeare's printer read 'limond' there as 'dam'd', the chances are that he substituted 'dambd' or 'damn'd' here for a word very like 'limond' in form. And, if we believe—as I think we should—that the Elizabethan top-level compositor is hardly less to be respected for his skills than his modern descendant, then the word can like 'limond' be allowed to be of some degree of difficulty.

The word that meets these conditions is 'limned', used elsewhere by Shakespeare in *As You Like It*, and as 'limning' in *Venus and Adonis*. If he here makes Iago call Cassio 'A fellow almost limned in a fair wife', the slight exoticism of 'limned' would only make the printer's misreading comprehensible. But the word would not seem alien to an Elizabethan, as it must sound to us, being now, of course, essentially archaic (though still fairly frequently met in its own art-historical context; London's Bond Street, for instance, offers a Limner Gallery, that deals in miniatures). For 'limning' is the art of miniature painting in particular, and of portrait painting in general, thought by many in the Renaissance to be the queen of all arts. For Iago in a sudden contemptuous, satirical paradox to diminish the promoted Cassio to the scale of an expensive miniature of a 'fair wife' gives his language an arrogant and precise expertise, a nailing clarity of image, that makes his insult all but unanswerable. Indeed, the very arrogance of the art itself may be the reason why Shakespeare makes his power-hungry and humiliated Iago take refuge in its vocabulary. For the two leading masters of the time, Hilliard and Oliver, were of course centred on the Court (though they did sometimes paint 'fair wives'), and Hilliard's descriptive defence of his art, the only recently published *Art of Limning* (written about 1600) is careful to establish the high status of the profession, its aristocratic ambience: 'None should medle with limning but gentlemen alone, for that it is a kind of gentill painting of lesse subiection than any other . . . and tendeth not to comon mens usse.'

If Iago does indeed call Cassio 'A fellow almost limned in a fair wife, | That never set a squadron in the field . . .', the insult holds the attention by a striking mixture of vitality with a curious deadliness: the man becomes a brilliant small flat empty doll. Though darkened, there is an interesting consonancy between this and the kind of praise given by Shakespeare to limning elsewhere. There is a tension in the celebration of *Venus and Adonis*:

> Looke when a Painter would surpasse the life,
> In limming out a well proportioned steed,
> His Art with Natures workmanship at strife,
> As if the dead the living should exceed . . .[4]

The same theme of art's conflicting and complex relation with the natural recurs in *As You Like It*, where Duke Senior, in recognizing and welcoming Orlando, tells the young man that he has his dead father's 'effigies' (monumental images) 'Most truly limn'd, and living in your face' (II. vii. 193–4). In both these passages Shakespeare's thinking is probably aided by a subdued pun on the word 'limbed'. The spelling 'limming' at line 290 of *Venus and Adonis*, where 'limning' is of course the primary meaning, is matched by the spelling 'lim' (for 'limb') at line 1067 of that poem. 'Limned' was

---

[4] *Venus and Adonis*, Ciij, 1593.

often actually spelled 'limbed' at the time; it may be that behind the Quarto *Othello*'s 'dambd' lies the Shakespearian spelling 'limbd'. This same shadow of another meaning enters Iago's line about the 'limning/limbing' of Cassio in a fair wife: where 'limbed', while contributing its own life, also complicates the situation into real indecency—it brings in what Sir Toby calls 'legges and thighes' to make the confused image more disturbing as well as more absurd.[5] The 'in' is both erotic and obscure. Cassio as a result not only beds with or in, but *becomes* the fair wife—a beast not with two backs but with one; the usual erotic frisson Iago aims at turns into a shiver of apprehension at Cassio's hermaphroditic sexual congress.

No actual member of an audience is going to catch more than the faintest hint of all this in Iago's 'limn'd/limb'd in a fair wife'—although it may be some uncertain sense of the disturbed, disturbing nature of the alternatives that has kept the editorial mind backed away from Iago's drift for so long. But it cannot be denied that the dark substance of these lines has its relevance to the play as a whole. *Othello* relates a man's loving to his seeing, and his seeing to his being—as he loves, so he becomes: and the metamorphoses of love emerge in their baser form through Iago's 'With her? On her: what you will'. This might be summarized by saying that in the tragedy 'limning' and 'limbing' break apart, and their disjunction generates terrible loose, rampant fantasies—under the clear surface of a simple story of sexual intrigue there start to spawn all the obscene, absurd varieties of sexual and emotional experience. To take one precise example: at least two distinguished actors (Olivier and MacLiammóir, as recorded in their memoirs) have toyed with the idea that Iago is in fact homosexually in love with Othello, and that his unhappy jealous passion triggers off the tragedy. This seems to me to give quite the wrong kind of interest to Iago's role, to render a false specificity in it. But the idea is not quite alien to the play; for it is a fact that *Othello*'s very first lines so

---

[5] And also perhaps to signal an opportunity for obscene gesture from the actor.

charge with curious undercurrents the exchanges of Iago and Roderigo, as to make them sound—if one wants—like a pair of lovers bitterly quarrelling, with the jilted and jealous Roderigo nagging the elsewhere-interested equally jealous Iago: 'Thou . . . hast had my purse, | As if the strings were thine'; 'Thou told'st me, | Thou did'st hold him in thy hate'; 'I follow him, to serve my turne upon him'.

Such effects are possible because the speakers here are both possessed by a violent, power-ridden, quasi-erotic rage that fails to 'limn' truly, to be in any true self-abnegating contemplation of its object. In some of the most terrible and powerful moments of a tragedy sometimes too easily called 'beautiful', this rampant and frightful imagination of love breaks free and holds sway. To all of these fantasies Iago gives the poisonous yet relaxedly commonplace style of 'limn'd/limb'd in a fair wife'. The grotesquerie of that line is not very far, for instance, from the black humour of the occasion (III. iii. 474–86) when Iago evokes for an appalled, entranced Othello the male mutuality of the pseudo-dream, in which calling 'sweet *Desdemona*' Cassio all but rapes Iago. The same muddy waters are plumbed again at IV. i, as Iago initiates a still partly truly innocent Othello into the unpleasing pleasures of their joint imagination of Cassio and Desdemona in bed together.

The sanity and depth of the tragedy lie in the way it can communicate that such images are both all too interesting to Othello, a form of experience that matures, yet all the time great agony to him. He never wholly loses his awareness that in losing Desdemona's true 'image' he will lose his peace for ever—an awareness that surfaces in moments of curious incoherent only half-conscious 'limning': 'My name that was as fresh | As Dians Visage, is now begrim'd and blacke | As mine owne face'; 'Turne thy complexion there: | Patience, thou young and Rose-lip'd Cherubin, | I heere looke grim as hell'. In the same way, Desdemona expresses her troubled but never lost love for Othello as insight or image. In her most directly sensual speech, when she pleads to be allowed Othello's physical company in marriage, she follows her allusion to what the Quarto calls her 'utmost pleasure' with the wonderfully abstract and metaphysical 'I saw *Othello*'s visage in his mind'. Her beautiful summary of a loving vision created of sympathy, an imaginative apprehension embodied and real, contrasts sharply with Iago's perverse notion of a Cassio 'almost limned' in some young woman.

Desdemona's definition of love has the kind of meditative complexity that we find in the Sonnets. There is a case still to be made that such textual difficulty as we find in the first edition of the Sonnets reflects less the incompetence of its printers than, often, an extreme difficulty in the thought expressed. It is a point of some interest that, exactly as in the Sonnets, defining Desdemona's 'metaphysic of the physical' seems to have caused Shakespeare such problems as to incur for the '*Othello*'s visage' lines at least five important textual cruces, places where the Quarto and the Folio so significantly differ as to make the probability of authorial revision seem a near-certainty. Here, too, textual problems may take us into the depths of a writer's work.

INDEX

Aaron (in *Titus Andronicus*) 188
Achilles (in *Troilus and Cressida*) 165–6, 170, 173–5, 179, 180–2, 185
Agamemnon (in *Troilus and Cressida*) 168, 170
Agate, James 11
Aguecheek, Sir Andrew (in *Twelfth Night*) 216–17, 219–20
Alceste (in *Le Misanthrope*) 125
All's Well That Ends Well 30, 49 n. 5, 65, 74, 163, 175
Alonso (in *The Tempest*) 130
Alton, R.E. 220 n. 3
Amadis de Gaulle 199
Amis, Kingsley 41, 183
Anne of Denmark, Queen 2
Antonio (in *The Merchant of Venice*) 40
Antonio (in *The Tempest*) 130
Antony and Cleopatra 84, 87
Arendt, Hannah 176 n. 2
Ariès, Phillippe 32 n. 17
Ariosto, Ludovico 206–7
Armado, Don Adriano de 204–5
Arnold, Matthew 75, 100–1
As You Like It 221–2
Auden, W. H. 140
Austen, Jane 46, 49
Avery, Gillian 29 n. 16

Baldick, R. 32 n. 17
Banquo (in *Macbeth*) 90
  his Ghost 84, 93–4
Barnardo (in *Hamlet*) 26, 134
Bates, Miss (in *Emma*) 49
Belch, Sir Toby (in *Twelfth Night*) 132, 216, 220–1
Benjamin, Walter 176–7
Bentivoglio, Hercole 203
Bergen, H. 178 n. 3
Bishopsgate, St Helen's 157
Bobadill (in *Every Man in His Humour*) 204

Bolingbroke (in *Richard II*) 123
Boswell, James 23
Bottom (in *A Midsummer Night's Dream*) 112
Boughner, Daniel C. 202 n. 10
Brabantio (in *Othello*) 44, 197
Bradamante (in *Orlando Furioso*) 207
Bradley, A. C. 17, 36, 80, 97
braggart soldier 202 ff.
Briggs, Julia 29 n. 16
Brooke, Arthur 111, 114, 116
Brown, Keith, 3 n. 1
Bryan, George 3
Burbage, Richard 15, 187
Burlador de Sevilla, El 200
Burrow, John 19–20

Calvin, John 100
Canetti, Elias 47
Cardenio 145
Cassio (in *Othello*) 41–2, 48, 205, 209, 212–15, 221–5
Caxton, William 165, 177–8
Cecil, Robert 191
Cervantes Saavedra, Miguel de 199
Chamberlain, John 23
Chandler, Raymond 11
Chapman, George 165, 181
Charlton, Kenneth 19 n. 8
Chaucer, Geoffrey 20, 49, 165–6, 172, 174
Chekhov, Anton 103
Christian IV of Denmark, King 2–4, 8
Christ's Teares (by Nashe) 140
Cicero 24
Cinthio, Giraldi 36, 39–40, 42, 45, 186–9, 200, 205, 207
Clarence (in *Richard III*) 119
Claudius (in *Hamlet*) 23–4, 26–7, 129–30, 141, 149, 155, 161–2
Clavijo, battle of 190
Collier, John Payne 151, 217
Collins, Wilkie 11

Columbus, Christopher 194
Commedia dell' arte 201
Compton-Burnett, Ivy 33
Cordelia (in *King Lear*) 60–2, 75–80
Coriolanus 84, 102, 157
Coriolanus 145
Coward, Sir Noel 218
Cressida (in *Troilus and cressida*) 37, 165–6, 171–5, 180–5
Crowds and Power (by Elias Canetti) 47
Cymbeline 38
Cyprus 43

Dante Alighieri 165
Dark Lady of the Sonnets 37
Decameron (by Boccaccio) 39
Desdemona (in *Othello*) 39, 41, 48–9, 55–6, 197, 205, 215, 224–5
Dickens, Charles 29
Dido, Queen of Carthage 178 n. 3
Divine Comedy, The 59
Dogs (in Shakespeare) 59–60
Don Quixote 199–201, 204
Donne, John 23–4, 89
Duke Senior (in *As You Like It*) 222
Dürer, Albrecht 128, 136

Earle, John 18 n. 7
Edgar (in *King Lear*) 60, 68–74, 78
Edmund (in *King Lear*) 61, 63, 121
Edwards, Philip 138–9, 142, 146–7, 150–1, 159–61
Elders and Betters (by Ivy Compton-Burnett) 33
Eliot, George 132
Eliot, T. S. 21, 129, 140
Elizabeth I, Queen 56, 132, 181
Elliott, J. H. 192 n. 4
Elliott, Mr (in Jane Austen's *Persuasion*) 46
Elsinore 1–7, 30
Emilia (in *Othello*) 39, 46, 57, 201
Emma (by Jane Austen) 49
Empson, Sir William 13
English History Plays 13, 129, 180
Essex, Robert Devereux, Earl of 56, 181

Falstaff (in 1 and 2 *Henry IV*) 25, 133
Faulconbridge, Bastard (in *King John*) 25
Ferdinand of Spain, King 194
Firbank, Ronald 49
Fitzgerald, Scott 41
Florio, John 141, 210
Fool (in *King Lear*) 66–8, 80–2
Ford, Ford Madox 41
Ford, Master (in *The Merry Wives of Windsor*) 16
Forman, Simon 84
Fortinbras (in *Hamlet*) 33, 127, 145–6
Frederik II of Denmark, King 3–5, 7–8
Frederiksborg 2, 4
Friar Lawrence (in *Romeo and Juliet*) 123
Furnivall, Frederick James 149

Gabiani, Vincenzo 203
Gaoler's Daughter (in *The Two Noble Kinsmen*) 153
Garrick, David 83
Gelosi, I (by Gabiani) 203
Geloso, Il (by Bentivoglio) 203–4
'Gerontion' (by T.S. Eliot) 21, 129
Gertrude (in *Hamlet*) 31, 153, 155, 160
Ghost (in *Hamlet*) 15, 153, 197
Glick, Thomas F. 192–3
Gloucester (in *King Lear*) 60, 63, 68–71
Goethe, Johann Wolfgang 28–9
Good Soldier, The (by Ford Madox Ford) 41
Goneril (in *King Lear*) 62, 66, 72–3
gravedigger (in *Hamlet*) 17, 31, 33, 120, 155, 161–2
Gray, Thomas 75
Great Expectations (by Dickens) 29
Great Gatsby, The (by Fitzgerald) 41
Greek New Comedy 201
Green Knight (in *Sir Gawain and the Green Knight*) 11
Guildenstern (in *Hamlet*) 25, 31, 158–9

INDEX

Gulliver (in *Gulliver's Travels*) 46

Hal, Prince (in 1 and 2 *Henry IV*) 133
Hall, Sir Peter 103
Hamlet 1–8, 11–33, 52, 59, 83, 124–36, 197
   Christmas in 14
   delay in 124–6
   Time in 128–35
   textual readings in 137–64
Hamlet, Prince 15–16, 24–5, 30–3, 43, 63, 124–30, 133, 136–8, 148–50
   age of 17–21
   indecorum of 149–50
Hamlet, the elder 53, 143–6, 148, 153
Harsnett, Samuel 71
Harvey, Gabriel 18
Hector (in *Troilus and Cressida*) 165–8, 173, 175, 182
Helen (in *Troilus and Cressida*) 184
Helsingør 3–4
Henry IV, 1 and 2; 16, 133
Henry IV, King 133
Henry V 131
Henry V 129
Henry VI 16, 129
Henryson, Robert 165, 174
Heraclitus 101, 127–8
Herrick, Marvin 203–4
Hieronimo (in *The Spanish Tragedy* by Kyd) 15
Hill, G. B. 22 n. 11
Hilliard, Nicholas 222
Hinman, Charlton 14 n. 1
Holmes, Martin 3 n. 1
Homer 165–6, 181
Honigmann, E. A. J. 3 n. 1
Horatio (in *Hamlet*) 15, 26, 34, 133, 137, 142–3, 145, 146–7, 150, 154–5
Hotspur (in 1 *Henry IV*) 145
Hunter, G. K. 196 n. 7

Iago (in *Othello*) 41, 46–51, 53–4, 56–7, 189–91, 195, 197, 205, 208, 210–16, 221–4
   as 'social' man 46–8
   Spanish name of 190
Ibsen, Henrik 103
Iliad (by Homer) 166
Importance of Being Earnest, The (by Wilde) 49
Innes, Michael 11
'Isabella' (by Keats) 145
Isabella of Spain, Queen 194
Isaiah 14
Italian learned comedy 201
Ivory Tower, The (by Henry James) 41
I Want It Now (by Kingsley Amis) 41

James, St (Santiago) 190–1
James VI and I, King 2, 85, 190
James, Henry 21, 29, 41, 181, 211
Jenkins, Harold 12 n. 1, 137, 139, 142–3, 146, 148, 150–2, 156, 158–61, 163
Jesus 19
Johnson, Dr Samuel 12, 22, 38, 75, 209
Jones, Eldred 196 n. 7
Jones, Emrys 38 n. 2, 97 n. 2, 201 n. 9, 204 n. 12
Jonson, Ben 204
Joyce, James 28–9
Juliet (in *Romeo and Juliet*) 16, 109, 115–18, 191

Kafka, Franz 170
Keats, John 145
Kempe, William 3
Kendrick, T. D. 190 n. 3
Kent (in *King Lear*) 60, 64, 68, 73–7
Kermode, Frank 188 n. 2
King Lear 13, 35, 58–82, 84
   love-test in 60–2
   love in 62–5
   Fool in 66–8
   Fool's disappearance from 81–2
   Edgar and Gloucester in 68–72
   Kent in 73–5
Kitely (in *Every Man in his Humour*) 204

Krogen 4, 7
Kronborg 1–7
Kyd, Thomas 15–16, 129, 140

Lady Capulet (in *Romeo and Juliet*) 109, 112, 116
Lady Macbeth 84, 92, 95, 97–8, 102–5
Laertes (in *Hamlet*), 24, 27, 31, 33–4, 132–3, 142, 150
Lafew (in *All's Well That Ends Well*) 74
Larkin, Philip 22
Laslett, Peter 116 n. 1
Lawrence, D. H. 174
Lazarillo de Tormes 200
Lea, Henry Charles 192 n. 4
Lear, King 15, 21, 63–5, 187, 217
Leskov, Nicholas 176
Levin, Harry 12 n. 1
Lord Chief Justice (in *2 Henry IV*) 133
Love's Labour's Lost 142, 204, 217
Love's Labour's Won 145
Lydgate, John 165, 177–8

Macbeth 59, 83–105, 190
  'success' in 95–8
  style of 89–90
Macbeth 21, 83–4, 88, 102–5
McClure, N. E. 24 n. 14
MacLiammóir, Michael 54, 223
Malcontent, The (by Marston) 144
Mallarmé, Stéphane 7
Malone, Edmond 152
Malvolio (in *Twelfth Night*) 217–19
Marcade (in *Love's Labour's Lost*) 142
Marcellus (in *Hamlet*) 15
Margaret, Queen (in *Richard III*) 121
Marlowe, Christopher 15–16, 178 n. 3
Marston, John 144
Mary I, Queen 199
Merchant of Venice, The 40–2, 187, 202
Mercutio (in *Romeo and Juliet*) 113, 123

Merry Wives of Windsor, The 9, 16, 38, 201
Midsummer Night's Dream, A 112, 114
Miller, E. H. 23
Miller, Jonathan 218
Milton, John 129
Le Misanthrope (by Molière) 125
Molière 125
Montaigne, Michel de 127, 140–1
Moors, Spanish 42, 191–207
Morocco, Duke of (in *The Merchant of Venice*) 188, 202
Mouse Trap, The (by Agatha Christie) 11
Much Ado About Nothing 38, 201

Nashe, Thomas 140, 201
Nestor (in *Troilus and Cressida*) 168, 170, 180–1, 185
North, Sir Thomas 148
Nosworthy, J. M. 138–9
Nurse (in *Romeo and Juliet*) 32, 109–13, 115–23

Old Wives' Tale, The (by Peele) 132
Olivia (in *Twelfth Night*) 133, 217–18
Olivier, Laurence (Lord Olivier) 6–7, 54, 82, 223
Ophelia (in *Hamlet*) 31–3, 135, 150–3, 155–8
Orlando (in *As You Like It*) 222
Orlando Furioso (by Ariosto) 206–7
Osric (in *Hamlet*) 34, 162–4
Otello (by Verdi) 35, 48, 206
Othello 35–59, 83, 186–207, 208–16, 221–5
  society in 45–52
  master and man in 47
  royalty in 51–3
  elopement in 53–4
  textual cruces in 208–16, 221–5
Othello (or 'the Moor') 15, 21, 37, 47–8, 50–1, 57–8, 187–8
  emergence of 43–4
  as 'Spanish' Moor 191 ff.
  colour of 197–8
Overbury, Sir Thomas 218

INDEX

Owen Wingrave (by Henry James) 29

Palamon (in *The Two Noble Kinsmen*) 152
Pandarus (in *Troilus and Cressida*) 180, 183–5
Paradise Lost 88 129
Parker, A. A. 200 n. 8
Parker, J. M. 157
Parolles (in *All's Well That Ends Well*) 49 n. 5, 74, 174
Patroclus (in *Troilus and Cressida*) 169–70, 179, 185
Paul IV, Pope 193
Peele, George 132
Petrarchanism 42
Philip II of Spain, King 199
Plato 135
Players (in *Hamlet*) 25–6, 153–4
Poe, Edgar Allan 11
Polacks (in *Hamlet*) 142–5
Polonius (in *Hamlet*) 27, 129, 149–52, 155, 158
Pope, Thomas 3
Portia (in *The Merchant of Venice*) 40, 195, 202
Powell, L. F. 22 n. 11
Pritchett, V. S. 11
Prospero (in *The Tempest*) 130–1
Puttenham, George 23
Pym, Barbara 49

Redgrave, Sir Michael 83
Regan (in *King Lear*) 62, 66, 72–3
Richard II 123
Richard III 119, 121
Richardson, Samuel 183
Ridley, M. R. 209
Roderigo (in *Othello*) 42, 189–91, 195–7, 208
  Spanish name of 189
Rogero (Ruggiero) in *Orlando Furioso*) 207
Roman comedy 201
Roman plays (by Shakespeare) 180
Romeo and Juliet 39, 109–23
  Lammas Eve in 114–15
Romeo 16, 112, 123, 191

Rosenborg 4
Rosencrantz (in *Hamlet*) 25, 31, 158–9
Rowe, Nicholas 217
Rymer, Thomas 38, 202, 205

S. Etienne 191–2
Sancho Panza (in *Don Quixote*) 200–1
Satan (in *Paradise Lost*) 88
Saxo Grammaticus 5
Seneca the Younger 15
Shakspere Allusion Book 15 n. 4
Shylock (in *The Merchant of Venice*) 40, 195, 198, 217
Simenon, Georges 11
Singer, Isaac Bashevis 63–4
Sir Thomas More 220
Sjogren, Gunnar 3 n. 1
Sonnets (by Shakespeare) 23, 37, 61, 65, 67, 79, 114, 225
Spencer, T. J. B. 154, 163
Stone, Lawrence 18 n. 6
Strindberg, August 103
Swift, Jonathan 181

Tate, Nahum 35
Tempest, The 130–1, 197
Theobald, Lewis 77
Thersites (in *Troilus and Cressida*) 165, 181–2, 184
Thomas, Sir Keith 18, 20
Thorello (in *Every Man in his Humour*) 204
Tilley, M. P. 210
Timon of Athens 60, 84
Tirso de Molina 200
Titus Andronicus 16
Tolstoy, Leo 67
Trial, The (by Kafka) 170–1
Troilus and Cressida 7, 37, 165–85
  plotlessness of 165–71
  reductiveness of 171–3
  betrayal in 173–5
  modernity of 176–81
  interest in the small 182–5
Troilus (in *Troilus and Cressida*) 165, 171–5, 180–5
Twelfth Night 133, 216–21

Two Gentlemen of Verona, The 60
Two Noble Kinsmen, The 152–3

Ulysses (by Joyce) 29
Ulysses (in *Troilus and Cressida*) 165–71, 174, 177, 179

Venice 40–3
Venus and Adonis 221–2
Verdi, Giuseppe 35–6, 206
Viola (in *Twelfth Night*) 133

Wain, John 188 n. 2
Waingrow, Marshall 22 n. 11
Waldock, A. J. A. 12 n. 2
Walker, Keith 22 n. 11
Way of the World, The (by Congreve) 49
Webster, John 89

Welles, Orson 144
Wells, Stanley 220 n. 3
Wilcock, G. D. 23 n. 13
Wilhelm Meister (by Goethe) 28–30
Williams, Kenneth 184
Wilson, John Dover 148, 154
Wings of the Dove, The (by Henry James) 41
Winter's Tale, The 38, 43
Witches (in *Macbeth*) 85, 101–2
Wolfit, Sir Donald 11
Wopsle, Mr (in Dickens's *Great Expectations*) 29
Wordsworth, William 28, 93
Wright, W. A. 217

Yeats, W. B. 32, 64
Yorick (in *Hamlet*) 17, 32, 120

Zohn, Harry 176 n. 2